Saving livelihoods saves lives

2018

Food and Agriculture Organization of the United Nations
Rome, 2019

Contents

Foreword

In recent years, the number of people experiencing hunger – both chronic and acute – has been alarmingly and persistently high. The annual State of Food Security and Nutrition in the World has repeatedly flagged global attention to the steady rise in the number of people experiencing hunger and malnutrition (815 million people in 2016 and 821 million in 2017), focusing on the role that conflict and climate change play in deepening hunger and vulnerability. At the same time, the annual Global Report on Food Crises has drawn attention to the growing number of people facing acute hunger.

Last year, 2018, was no exception. Some 113 million people in 53 countries suffered from acute hunger, according to the Global Report. That is 113 million girls, boys, men and women, old and young, who were unable to access enough food and required humanitarian assistance.

Much of this hunger is driven by stresses – conflict, climate and economic shocks – that have disrupted livelihoods and left people unable to meet their needs. However, for the most part, the hunger is the result of a constant erosion of livelihoods and food systems – as a result of climate change, conflict and political instability, environmental degradation and repeated shocks.

For the Food and Agriculture Organization of the United Nations (FAO), building resilient agriculture-based livelihoods and food systems is at the core of its efforts to fight acute hunger and avert food crises.

We know how critical humanitarian assistance is – for example in Yemen, where the scale of hunger and human suffering is staggering but which would be considerably worse without the provision of humanitarian assistance.

At the same time, it is clear that humanitarian assistance on its own is not enough to win the battle against acute hunger. That is why FAO's humanitarian work is firmly embedded in a foundation of building resilience.

This was demonstrated in 2018, when our work extended from immediate humanitarian response to protect lives and livelihoods in some of the most complex contexts in the world, including South Sudan and Yemen, to addressing the vulnerability of pastoral populations and facilitating the development of livestock feed balances in the Horn of Africa, to supporting disaster risk reduction efforts from the Philippines to Central America.

This is the true strength of FAO's resilience programme – using timely information and analysis, a strong evidence base and building on experiences in different contexts to safeguard and build resilient agriculture-based livelihoods and food systems even in times of crisis.

Last year, FAO's resilience programme reached 25 million people through a combination of short- and medium-term actions intended to ensure their continued access to food, reduce acute hunger and build resilience.

However, as always, this number represents just a portion of those in need. In 2018, 44 percent of the funds requested under various appeals was received, meaning that millions of farmers, herders, fishers and foresters remained without critical livelihoods assistance. In 2019, we will continue working with our partners to ensure we make the best use of limited resources and reduce the number of people that need our assistance by focusing much more on addressing the root causes of their vulnerability.

This publication offers us an opportunity to reflect on some of our achievements over the past year and identify how we can do better in the future. It is not an exhaustive list of all of FAO's resilience work, but rather an overview of what we can achieve and how much more there is to be done.

No one agency can tackle food crises on its own. That is why FAO is actively engaging with a wide range of partners at country, regional and global levels. In 2018, this included strengthening our partnership with the other two Rome-based Agencies (the International Fund for Agricultural Development [IFAD] and the World Food Programme [WFP]), as well as with other agencies, such as the International Organization for Migration (IOM), the Office of the United Nations High Commissioner for Refugees (UNHCR), the United Nations Educational, Scientific and Cultural Organization (UNESCO), the United Nations Children's Fund (UNICEF), the World Health Organization (WHO) and the World Organisation for Animal Health (OIE), among others. We have also invested in our relationship with regional organizations like the Central American Integration System (SICA), the Permanent Interstate Committee for Drought Control in the Sahel (CILSS), the Intergovernmental Authority for Development (IGAD), and the Southern African Development Committee (SADC), particularly in the context of food security information and analysis and resilience measurement.

In 2018, with considerable support from the European Union, we moved ahead in operationalizing the Global Network Against Food Crises, which focuses on preventing and addressing crises, bringing together actors across the humanitarian-development-peace nexus, recognizing that it is only by tackling the root causes of hunger that we can avert food crises in the future.

In addition, greater investment in multi-year resilience programmes is critical. In 2018, we saw significant progress in developing and implementing such programmes, but they must become the norm and not the exception. Unless we invest at scale in building resilience – beyond a handful of projects that target 20 000 people here, 100 000 people there – we will not make significant gains in reducing acute hunger. That is the reality.

©FAO/L. Tato

José Graziano da Silva
FAO Director-General

FAO assists people to strengthen their livelihoods to withstand crises, which saves lives, and reduces humanitarian needs and costs by a significant margin.

Snapshot

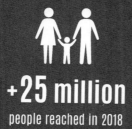

+25 million
people reached in 2018

+60 million
livestock treated/vaccinated

+7 100
professionals trained to prevent, detect
and respond to disease threats

Highlights

Increasing food production

In 2018, FAO provided almost 430 000 families in **South Sudan** with 4 800 tonnes of seeds in time for the main planting season, resulting in a total yield of almost 273 000 tonnes of cereals (maize and sorghum), representing nearly one-third of all cereal production in the country last year.

In **the Philippines**, 184 tonnes of rice seeds distributed produced over 2 700 tonnes of milled rice that could feed more than 24 000 people in one year.

In **the Central African Republic**, 1 600 tonnes of cereal seeds produced over 20 000 tonnes of cereals for 250 000 people.

Enhancing livelihoods

In **Somalia**, USD 27 million in cash transfers was provided – including through cash-for-work programmes that rehabilitated a total of 325 agricultural infrastructures.

In **Haiti**, 105.3 tonnes of crop seeds, 6.88 million planting materials and 240 kg of vegetable seeds were provided to drought-hit families, who each harvested at least 850 kg of cereals, legumes and tubers.

Preventing and protecting

In **Iraq**, mortality rates were reduced by 95 percent among livestock in Nineveh governorate thanks to FAO's vaccination programme, while in **Burundi**, mortality rates were reduced by 100 percent among small ruminants vaccinated against *peste des petits ruminants* (PPR).

Contributions received in 2018

for resilience programmes, including emergencies

USD 617 million
received

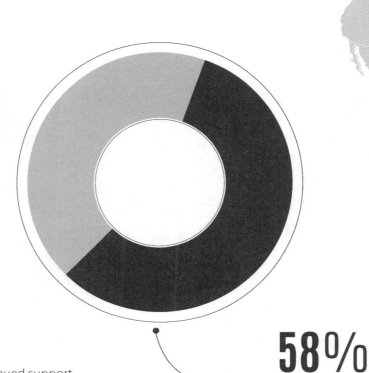

Thanks to the continued support of our resource partners, resource mobilization increased by

USD 45 million
from last year

58%
went to
10 countries

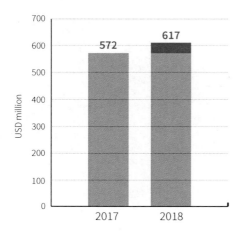

1	**Somalia** USD 121.3 million	**6**	**Chad** USD 16.4 million
2	**South Sudan** USD 77.9 million	**7**	**Afghanistan** USD 14.4 million
3	**Yemen** USD 38.5 million	**8**	**Syrian Arab Republic** USD 13.2 million
4	**Pakistan** USD 31.0 million	**9**	**Uganda** USD 13.1 million
5	**Nigeria** USD 20.1 million	**10**	**Burkina Faso** USD 12.0 million

Resource partners
invested in

70
countries

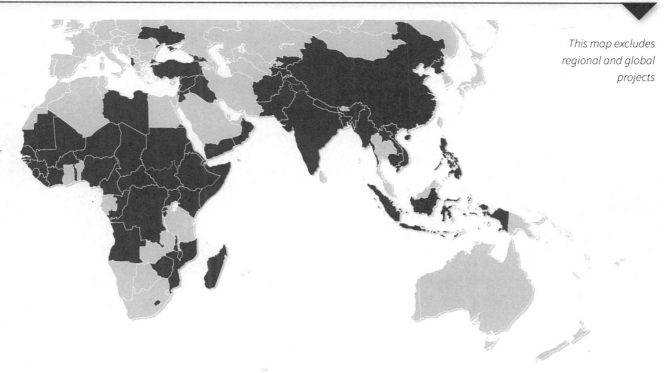

This map excludes regional and global projects

Top 15
resource partners in 2018

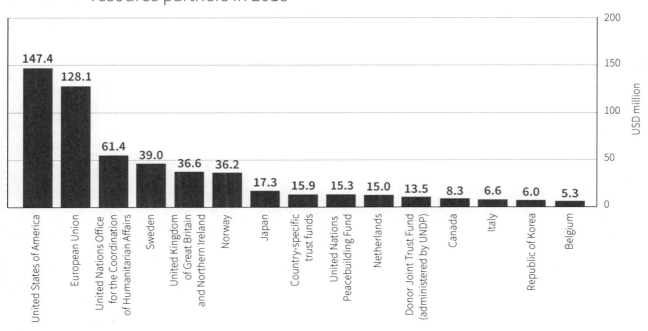

	USD million
United States of America	147.4
European Union	128.1
United Nations Office for the Coordination of Humanitarian Affairs	61.4
Sweden	39.0
United Kingdom of Great Britain and Northern Ireland	36.6
Norway	36.2
Japan	17.3
Country-specific trust funds	15.9
United Nations Peacebuilding Fund	15.3
Netherlands	15.0
Donor Joint Trust Fund (administered by UNDP)	13.5
Canada	8.3
Italy	6.6
Republic of Korea	6.0
Belgium	5.3

Anticipating and preparing for crises

A shock does not inevitably lead to a crisis. When families, communities, and local and national authorities have the most accurate information on a potential risk, they can prepare for and even mitigate the worst effects of a threat so that it never becomes a crisis.

Resilient livelihoods are built on a solid foundation of accurate and timely information that directs preparedness, prevention, mitigation and response actions as well as identifies risks and vulnerability, through food security analysis, needs assessments and resilience measurement.

FAO plays a leading role in global, regional, national and local level early warning systems and analyses, including as one of the main partners of the Integrated Food Security Phase Classification (IPC). In Yemen, for example, FAO coordinates the IPC process together with more than 20 organizations – United Nations agencies, Non-governmental Organizations (NGOs), government ministries and parastatals, private sector and academia. In 2018, huge progress was made in further refining the IPC process in Yemen and getting closer to those in need. The latest IPC analysis (December 2018) covered 333 districts – a considerable improvement from the last analysis (March 2017), which covered a larger administrative (governorate) area. This was a major undertaking, particularly given the difficult conditions in Yemen, and ensures that all stakeholders have a much clearer picture of the situation at district level and can therefore tailor their interventions to provide the most appropriate support to those most in need. Last year, the IPC technical working group co-led by FAO conducted seven rounds of acute food insecurity analysis, with over 70 experts participating, representing different organizations and sectors.

Even in the midst of a major humanitarian crisis, like Yemen, it is possible to improve early warning and information for other threats. For example, in 2018, FAO collected, analysed and reported market price information, food imports and foreign exchange to inform updating of the cost of the minimum food basket and facilitate humanitarian programming. FAO also regularly monitors other risks including plant pests and natural hazards.

©FAO/S. Ahmed

FAO participation in needs assessments in 2018

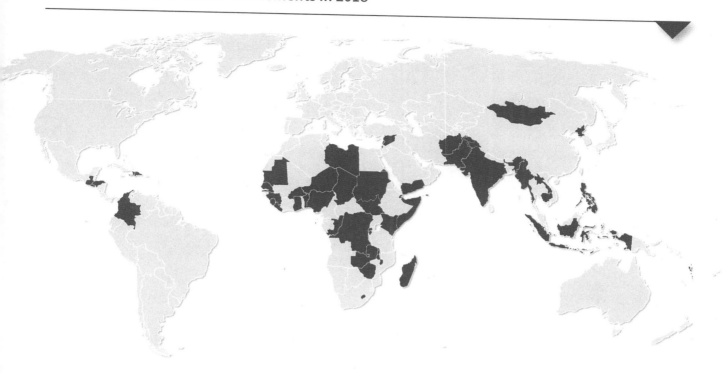

When alerts point to a deteriorating situation in a country, or when a disaster does strike, FAO often leads or participates in needs assessments to determine the impact of a crisis on agricultural livelihoods and food security and guide future actions. In 2018, FAO was part of 74 needs assessments in 52 countries, from the Dominican Republic to Zimbabwe, Tonga to Mauritania.

One example is Colombia, where the food security and nutrition status of vulnerable rural communities deteriorated significantly. In July 2018, FAO, WFP and UNICEF conducted a joint needs assessment to assess the impact of the migratory wave from Venezuela in rural areas of La Guajira Department. Through needs assessments, FAO was able to flag deteriorating food security and respond quickly to protect lives and livelihoods.

In Pakistan, FAO participated in a Drought Needs Assessment in Sindh province in October 2018, which found that the prevalence of food insecurity based on the Food Insecurity Experience Scale (FIES) is high, and cultivation areas and crop production for wheat, rice, cotton, cluster beans, millet and pulses was significantly lower in the 2017-2018 agricultural seasons, compared with 2016-17. The livestock sub-sector has also been adversely affected, with high death rates reported for livestock. Based on the results,

the Government and other decision/policy makers and partners were able to prioritize actions in relevant sectors to address immediate needs and increase future resilience to drought.

In 2018, FAO, WFP and UNICEF also sounded the alarm at the deteriorating food security of particularly pastoral and agropastoral communities in the Sahel, with a 60-percent rise in the number of people at risk of food insecurity as a result of drought, high food prices, persistent conflict and market distortions in parts of the region.

In many countries already facing serious food crises, local food systems face the further threat of animal and plant pests and diseases. In 2018, for example, FAO helped small-scale producers around the world to reduce the potential threats of pests like fall armyworm (FAW) and diseases like Rift Valley fever (RVF).

©FAO/A. G. Farran

In South Sudan, when an outbreak of RVF was recorded in Eastern Lakes State in 2018, among other activities, livestock and animal health experts from FAO and the Ministry of Livestock and Fisheries trained cattle keepers and breeders on how to identify the signs of the disease and what to do in case of RVF detection. These sessions were recorded and radio programme campaigns were aired in Tonj, Yirol and Rumbek, targeting livestock owners and enabling them to quickly detect and react to further outbreaks of the disease.

Simulation exercises are a key part of efforts to prepare for potential infectious disease outbreaks, which can have severe implications for food security,

nutrition and livelihoods. Given the continued threat of highly pathogenic avian influenza (HPAI) in Africa and beyond, FAO continued to work with local and national authorities in the region to help them prepare for a potential outbreak and address any potential gaps in their response. In March and August 2018, more than 60 field veterinarians from ten provinces of the Democratic Republic of the Congo participated in a one-day simulation to better recognize the clinical signs of avian influenza. In October, an avian influenza simulation exercise in Chad helped 50 officers from the Ministries of Livestock, Environment and Public Health to develop One Health Contingency Plans.

Enhancing national preparedness
in Kenya

In Kenya, FAO has been at the forefront in strengthening the capacity of government, non-state actors, vulnerable communities and households to mitigate, predict and/or effectively respond to disasters impacting agricultural livelihoods. This includes the Government's endorsement of the Predictive Livestock Early Warning System (PLEWS) through the National Drought Management Authority for implementation. The Government has also adopted the FAO-developed Resilience Index Measurement and Analysis (RIMA-II) tool for the baseline survey of the Kenya climate-smart agriculture programme, with FAO providing technical support on RIMA analysis. In addition, FAO has facilitated the testing and validation of the preparedness capability of Kenya's animal and public health systems to prevent, detect and respond to potential outbreaks of HPAI and RVF, leading to the revision of the relevant national preparedness and contingency plans. FAO has also helped to strengthen the

early warning capacity and effective management of the RVF outbreak that occurred from June to July 2018 through a range of interventions including: sentinel herd monitoring with collection of 568 samples in six counties; provision of laboratory supplies for the diagnosis of RVF; provision of technical support/advice for the management of the RVF outbreak by actively participating in the national RVF technical working group; training 16 county disease surveillance/reporting officers on management of RVF; and sensitization of 20 county-based media personnel from 15 counties on RVF for factual reporting on risks and outbreaks.

FAO has also helped to enhance the timely and real-time sharing of laboratory diagnostic results and other related information through establishment of a laboratory information management system known as SILAB, which interconnects the national reference laboratory and the regional veterinary laboratories. Animal

disease reporting rates have improved from 25 to 72 percent in seven counties following the introduction of a mobile-based real-time disease reporting tool known as Kenya Livestock and Wildlife Syndromic Surveillance and subsequent training of 269 frontline disease surveillance/reporting officers on the use and application of the tool. By adopting this system and improving disease reporting, sensitivity of the national animal health surveillance systems will be increased, thereby strengthening the country's capacity to early detect and rapidly respond to outbreaks of livestock and wildlife diseases.

In addition, FAO contributed to enhancing the diagnostic capacity, quality system as well as biosafety and biosecurity procedures of the Central Veterinary Laboratory. FAO also selected regional veterinary investigation laboratories to enable the country to detect rapidly and accurately priority pathogens, including anti-microbial resistance.

The In-Service Applied Veterinary Epidemiology Training programme was launched in October 2018, with pilot trainings and needs assessments taking place in 14 countries across Africa. Throughout the year, eight missions to Africa using FAO's Surveillance Evaluation Tool recommended national improvements for spotting infectious diseases.

©FAO/G. Lamielle

"This in-service training for veterinary epidemiologists is a good model for future sustainability as once we have built in the momentum together, it can be led and expanded by local and continental veterinary institutions."

Juan Lubroth, FAO Chief Veterinary Officer

Preparing for potential avian influenza outbreaks
in Mali and Senegal

Mali and Senegal are under threat from HPAI infection. In order to help key local actors to prepare for a potential outbreak, FAO carried out a cross-border simulation exercise in May 2018. The H5N1 subtype of HPAI has caused severe damage to the poultry industry in western and central Africa in recent years, with implications for nutrition, as poultry is often the cheapest and most widely available source of protein for many of the most vulnerable families. Humans are also at risk of catching the infection, which can be fatal.

The two-day simulation involved 60 representatives from government, local and regional police, regional veterinary services and regional authorities. During the training, the participants did a test-run of their national contingency plans in the border communes of Kidira in Senegal and Diboli in Mali. Six working groups responded to different hypothetical scenarios – from sampling mortalities in a hen house, to containing a live bird market with positive test results, to raising awareness of risky behaviours among local populations. Participants tested their practices and procedures for biosafety and biosecurity, testing and diagnosis, communication and decision-making.

In the last three years, new outbreaks of HPAI subtypes H5N1 and H5N8 have been recorded in six countries in West and Central Africa. Although Mali and Senegal have not recorded any cases of HPAI infection, the threat is present due to the affected neighbouring countries. Joining forces in preparedness helps to improve their capacity to prevent and respond to HPAI outbreaks in the border areas. As a result of the simulation, roles and responsibilities were reinforced and difficulties implementing biosecurity measures were highlighted. New training and guidelines were subsequently developed to enhance a One Health approach to preparedness – involving animal, public and environmental health stakeholders.

"After the training, the maize harvest was good. If I had not gotten the training my maize would all be destroyed by the insects."

Alice Sunday Serafino Lasu,
Farmer from Juba, South Sudan

Plant pests and diseases are also a major threat to people's livelihoods and outbreaks can exacerbate already high levels of food insecurity. For example, in South Sudan, smallholder farmers were forced to cope with an outbreak of FAW in 2018, when millions were already experiencing acute hunger. In order to avert a further crisis, FAO worked closely with the Ministry of Agriculture and Food Security at the state and national levels to assess the severity and spread of the FAW infestation and initiate measures to assist farmers in combatting the pest. This also involved a strong partnership with WFP, by supporting farmers that had been affected by the pest to recover by providing them with food aid and social protection activities, allowing them to re-establish their livelihoods.

FAO trained 104 frontline village facilitators from different counties in South Sudan who provided hands-on advice to farmers, and monitored and assessed the impact of the pest across the country at community level. Community facilitators were given tablets to monitor the infestation and public service announcements were played on radio stations across the country to educate farmers. FAO's interventions helped affected communities to avoid total crop losses in 2018.

FAO, together with Penn State University, developed an innovative new mobile phone app to help farmers, communities and extension workers identify FAW and collect and record information on damages caused by the pest. The app, which uses artificial intelligence to diagnose if a maize plant is infected by FAW, was launched in March 2018 across Africa and translated in 13 languages. Since its release, more than 55 000 reports had been received through the app from 40 countries.

In addition, FAO has developed the FAWRisk-Map, which incorporates diverse socio-economic and agro-ecological data so that responders can visualize where the underlying risk of household food insecurity due to FAW is highest. The tool consists of a number of layers allowing users to disaggregate risk into its constituent parts. By highlighting potential "hotspots", the tool is intended to assist decision-makers in prioritizing and preparing for early action in targeted areas.

FAO helps Kenyan farmers rescue harvest from fall armyworm

in Kenya

Agnes Waithira Mull, is a smallholder farmer in Embu county in central Kenya. She and her husband lost most of their last crop due to FAW. Thanks to FAO training in FAW control, however, they are better able to protect their current crop. "Now that we know how to deal with the infestation, our losses will be smaller," Agnes says.

During the 2017 and 2018 short rains season, FAO initiated a pilot project, where specially trained 'field scouts' were deployed to visit smallholder farmers and assist them in manual FAW control, twice a week for six weeks in Embu and Bungoma Counties. The method used was mechanical control – identifying the eggs and larvae and then destroying them by hand.

Agnes and her husband, Robert Nurithi Nthiga, became part of this initiative. As a result, they did better than many of their neighbours - those who did no mechanical control often lost much of their crop to FAW.

The couple have one acre of land, but it is divided into several plots. "We were only able to do the control on a quarter of the land, where we got four 90-kg bags of maize. Without the FAW, we could have harvested from the whole area, and got at least 15 bags," Agnes says.

These days, without the support of a scout, she and her husband inspect their plots for the insect as often as twice a day. "It's a lot more work than before, but checking on the crop is a farmer's job," she says. The mother of three children, the youngest less than a year old, Agnes is busier than ever before.

A recent evaluation confirmed the positive sentiments from project participants and continued use of

©FAO/S.Lazaro

mechanical control beyond the end of the project. In particular, 80 percent of participants who used mechanical control during the first phase of the project (2017) continued it without field scout support during the following 2018 long rains season, and 58 percent indicated that they preferred mechanical control to other practices. The assessment also found that the programme benefited the community beyond the direct project participants as each participant told six other community members about the mechanical control (mean value).

Knowing when shocks are likely and who is vulnerable means that FAO can work with at-risk communities and partners to prepare for crisis by safeguarding local livelihoods, reducing the potential impact of a shock and the likelihood of families adopting negative coping mechanisms, like selling important assets, taking on debts or limiting their food consumption.

Through the Regional Committee of Hydraulic Resources of the Central American Integrated System (SICA), in 2018, FAO consolidated an Agricultural Drought Surveillance System for Central America, which allows countries to monitor the behaviour of basic grains that are essential for food security and nutrition, as well as for local economies. The system operates in Belize, Costa Rica, El Salvador, Guatemala, Honduras, Nicaragua and Panama, with the satellite information system detecting agricultural areas with a high probability of experiencing water stress (drought). The Regional Committee of Hydraulic Resources brings together the meteorological services of SICA members and, with the support of the World Meteorological Organization and FAO, generates data to alert members to risks that may affect crops and therefore food security.

The Agricultural Drought Surveillance System is critical, especially given the increasing number of people living in marginal areas and that are vulnerable to drought in the region. In particular in the Dry Corridor, which extends from Mexico to Panama and encompasses a highly vulnerable rural population, where a lower crop yield due to water stress can significantly impact people's access to food. The System provides timely information so that Ministries of Agriculture can understand and react to potential risks. In the future, the System should be closely linked to early action plans to mitigate the impacts of extreme drought events.

Assessing natural resource degradation in refugee settlements
to increase resilience and ensure energy access

By March 2018, over 1 million South Sudanese refugees and asylum-seekers had fled to Uganda and Kenya. Adding to existing high rates of resource degradation, their arrival has had a range of environmental impacts including land degradation and woodland loss, resulting in inadequate access to energy for cooking and competition for natural resources. Given the combined pressures from the existing population and new arrivals, there was a clear need to develop strategies for sustainable energy access and forest resource management targeting both refugee and host communities.

In 2018, FAO conducted two assessments to determine the environmental impacts of the South Sudan refugee influxes in northern Uganda and northeastern Kenya, with a focus on forest resources, and propose practical intervention options to mitigate pressure on the environment, support energy access and contribute to livelihoods resilience of refugees and host communities.

Among others, the assessment determined that the refugee influx from South Sudan to northern Uganda has led to an increase in the rate of degradation and tree loss both inside the refugee settlements and around their boundaries, with accelerated land cover changes in bushland and woodland. Total cooking fuel demand in 14 of these settlements is about five times the quantity of tree growth within the settlements and the 5-km buffer, which could result in annual biomass loss of about 10 percent.

The assessment recommended a range of costed interventions and additional measures, including rehabilitation of degraded forests using both natural and assisted regeneration; establishment of woodlots to increase supply of woodfuel and other products such as building poles, fruits and fodder; development of agroforestry systems on household plots and farmland; and enhancement of energy efficiency.

These would improve environmental management, ensure access to woodfuel resources for both refugee and host communities, and contribute to building livelihood resilience.

Preparing herder communities against anticipated shocks
in Palestine

Agriculture-dependent communities in Area C of the West Bank, particularly Bedouin and herder communities, face substantial challenges in accessing essential natural resources, like water and grazing lands, while the high cost of fodder means it is often far beyond their reach. Area C constitutes over 60 percent of the West Bank and access to resources for farming and herding is severely restricted.

While herders have historically adapted to their environment with well-tailored coping strategies; over 60 years of conflict has slowly undermined their adaptive capacity, rendering their livelihoods increasingly fragile and unsustainable.

In 2018, FAO helped over 6 000 herding families to address these crucial constraints by rehabilitating, and in some cases constructing, 178 water cisterns that harvest rainwater for livestock. Each cistern has a storage capacity of between 70 and 200 m³, with a total capacity of over 25 175 m³. These cisterns have improved families' livestock production, as well as generated over 890 temporary jobs through construction and rehabilitation works and eased the burden on women in these communities, who used to walk long distances to fetch water for livestock and for daily household chores.

At the same time, in order to ensure greater access to feed, FAO provided almost 320 tonnes of drought-tolerant seeds for fodder crops, enabling herders to cultivate up to 2 538 ha of grazing land, producing fodder of a total value exceeding USD 2.6 million for a cost of just USD 234 814. While normal fodder varieties produce about 50 kg of dry matter, the drought-tolerant seed provided to farmers can produce up to 400 kg of fodder, meaning herders are less dependent on imported fodder and price volatility in local markets.

Finally, in an effort to protect valuable livestock assets before and during the lambing season, FAO distributed vitamin and mineral supplements and antimicrobials to herders, while animal shelters were equipped against anticipated harsh winter conditions. Some 380 000 m² of plastic sheds were distributed to herders in Palestine, helping to reduce animal diseases and mortality. Herders' access to markets was strengthened through the renovation of a livestock market that includes a veterinary service centre.

In early 2018, FAO organized a training for 35 officials from the Pakistan Metrological Department (PMD) from Punjab, Sindh and Balochistan provinces, on the use of the global Agriculture Stress Index System (ASIS) that PMD uses to detect areas with a higher likelihood of prolonged dry periods and drought. The ASIS system has already started generating drought alerts and will help PMD to forecast the potential drought levels in various districts.

In Bangladesh, in anticipation of the monsoon season and its potential impact on already extremely vulnerable displaced and host communities, FAO worked with local and national forest authorities to support land stabilization and reforestation efforts to reduce the risks of landslides.

While advocating for greater global investment in early action, FAO also reinforced internal links between early warnings and early action on the ground, which has been critical to minimize the impact of crises and protect lives and livelihoods, assisting people at risk in Mongolia, the Sudan, and across southern Africa during 2018.

©FAO/M. Longari

Early Warning
Early Action

In 2018, the international community collectively made great strides towards further advancing anticipatory approaches to disasters as a much-needed complement to traditional humanitarian response. Early action – lessening the impact of disasters on people through early warning-triggered mitigation and prevention measures – has become increasingly mainstreamed with a view to transforming the way disasters are managed in the coming years.

FAO, with its Early Warning Early Action initiative, is very much at the centre of these developments, contributing through its ground-breaking country, regional and global work. Early actions for a variety of risks and agriculture sectors were implemented across Africa, Asia and Latin America throughout 2018. The results are promising, pointing to significant cost-effectiveness of early action, both in terms of saving key livelihood assets and in terms of lessening humanitarian response costs. Protecting people and livelihoods ahead of shocks helps build resilience to future ones.

Changing the way we manage disasters at the global level

FAO has continued to invest in early warning early action capacities globally, supporting a vision for a system-wide shift towards increased anticipatory action.

FAO's quarterly Early Warning Early Action report on food security and agriculture is a landmark forward-looking analysis that helps the international community to prioritize risks and encourages systematic early action. In 2018, the report was redesigned to further strengthen its quality and user-friendliness, with a view to potentially becoming a key collective statement on future risks to agriculture and food security.

Another critical milestone in the international dialogue on early action was the launch, in March 2018, of the first Inter-Agency framework for early action – the Standard Operating Procedures (SOPs) for Early Action to El Niño/La Niña events. Based on the lessons learned from the 2015/16 devastating El Niño event, the SOPs – co-led by FAO – represent an important step towards envisioning how different actors can work together to share early warning information, and initiate early actions. With an El Niño event looming in late 2018, the SOPs were activated, providing the international community with a common understanding of the potential impacts and high risk areas, calling for close monitoring and planning.

©FAO/A. Adil

Early Warning Early Action approach

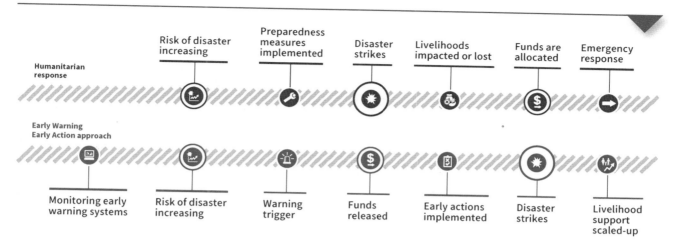

Humanitarian response

- Risk of disaster increasing
- Preparedness measures implemented
- Disaster strikes
- Livelihoods impacted or lost
- Funds allocated
- Emergency response

Early Warning Early Action approach

- Monitoring early warning systems
- Risk of disaster increasing
- Warning trigger
- Funds released
- Early actions implemented
- Disaster strikes
- Livelihood support scaled-up

Acting early to save lives and livelihoods at regional and country levels

Early actions were implemented in different contexts throughout 2018, highlighting the flexibility with which this approach can be applied to different geographical areas, livelihood sectors and hazards. Triggered by warnings of an escalating drought in the Sahel, early actions were implemented to protect livestock through vaccinations and the distribution of feed and nutrient supplies.

Exploring how early action can be applied to human-induced hazards, a foreseen increase in the influx of migrants from Venezuela prompted early actions to boost household food production in the Colombian border areas. In late 2018, warnings of El Niño-induced rainfall deficits prompted mitigation activities in the Philippines as well as across southern Africa to protect livelihood assets and food production, curbing future food insecurity.

Protecting livelihoods against future shocks through early action has proven to be highly effective – whether in response to drought, severe winter, animal diseases or migration. In a landmark effort, FAO collected empirical evidence on the Return on Investment of acting early. In 2018, early actions were assessed across three countries: the Sudan, to protect livestock ahead of localized dry spells, Madagascar to support small-scale farmers ahead of drought, and Mongolia to support vulnerable herders through

©FAO/C. Jones

a severe winter season. These empirical studies provide a critical snapshot into the value for money of acting before an anticipated crisis has become a humanitarian disaster.

Benefits of Early Warning Early Action

Protecting livestock against the *dzud* in **Mongolia**

What was the return on investment?

USD 1 ⇒ USD 7.1

USD 2 008

of avoided losses and added benefits for each household

In 2017/18 a localized *dzud* was forecasted in Mongolia. The phenomenon is characterized by a summer of very high temperatures and little rainfall followed by the harshest of winters. The heat degrades the pasture and limits fodder growth. Then the snow and frozen ground stop livestock from reaching even the scant remaining grazing. Ganbaatar Sodnom-ish was especially vulnerable as he has fewer livestock to take to market – especially when prices tumble, as everyone else is desperate to sell. He benefited from FAO support when warning signs pointed to upcoming *dzud*.

"Because our animals survived the dzud, *we could use them as a guarantee for a loan. We used this loan to buy a motorbike, which is perfect for us in the countryside. I save a lot of time taking my children to school and getting into the town centre."*

Ganbaatar Sodnom-ish hasn't paid off the loan he took out to buy a motorbike yet, but the interest rate is reasonable at 2.1 percent. As he lost no animals when he received food and nutrients from FAO's Early Warning Early Action programme, Sodnom-ish hasn't borrowed more money to keep his herds – and the income they generate – at a stable level. He's very happy with what he has achieved. Sodnom-ish supports his family of five with his herd. He received enough feed for most of his animals for 60 days and a cash payment for selling meat after participating in FAO's destocking programme and sees a positive future for himself and his family.

©FAO/P. Khangaikhuu

Coping with drought in **the Sudan**

 **What was the return on investment?**

USD 1 ➡ USD 6.7

USD 431

of avoided losses and added benefits for each household

Khalda Mohammed Ibrahim has a small herd of goats and sheep, providing meat and dairy for her family, especially her 1-year old daughter. Milk is vital for children's nutrition. Just 0.5 litres/day gives a 5-year old 25 percent of the calories and 65 percent of the protein they need.

From August 2017, worrying signs emerged in Kassala State, where Khalda lives – extended dry spells, unusual livestock movement and high market prices. Based on this, FAO implemented early actions to support households like Khalda's,

who rely solely on pastoralism. Khalda was happy with the timely livestock feed and nutritional supplements she received. These meant she was able to keep her livestock healthy and their milk production increased, when animals are typically at their weakest.

"With this help our livestock were healthy… and made more milk. We could feed the children and ourselves and sometimes we provide the neighbours with milk."

Easing the impact of drought in **Madagascar**

 What was the return on investment?

USD 1 ➡ USD 2.5

USD 78

of avoided losses and added benefits for each household

Zarafonomeny relies on a mix of farming activities to provide for herself and her four children. She also collects firewood and harvests braid – a local green-leafy crop – to sell at the market.

©FAO/H. Razafindralambo

Southern Madagascar, where Zarafonomeny lives, has faced cumulative and intense droughts, resulting in three failed agricultural seasons. In 2017/18, when forecasts pointed to a fourth drought, FAO was able to act early before the drought reached its peak.

FAO supported Zarafonomeny with short-cycle seeds – including groundnuts and vegetable – to help her cope despite the drought.

"With the seeds and help for my water supply we were able to grow more vegetables and also different kinds. We ate some but also took them with us to the local market to sell with the firewood. The money I got from this gave me room to cover some of our daily needs, such as buying rice, oil or soap".

Responding fast to crises

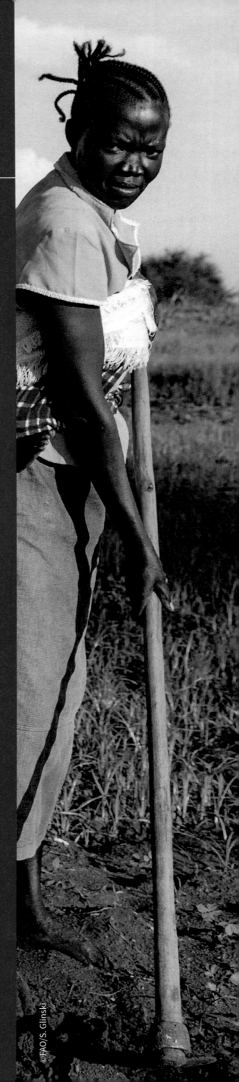

When conflict, extreme weather events, natural disasters, and other threats devastate rural livelihoods and push people into acute food insecurity, FAO is on the ground providing critical productive inputs, like seeds or fishing gear, to rapidly restart food production; or vaccinating pastoralists' animals on a massive scale to safeguard their livelihoods.

In 2018, FAO continued to provide critical support to rapidly restore livelihoods – from rapid response to cash and agricultural inputs, as well as further strengthening its support to the global Food Security Cluster and other country-level coordination mechanisms.

In addition, FAO has ongoing agreements with standby partners that enable the Organization to rapidly deploy experts and scale up activities at a moment's notice, fill urgent staffing gaps and ensure smooth response to crises. In 2018, a total of 50 experts were deployed by standby partners for 9 220 days, representing a 55 percent increase from 2017 and a 150 percent increase from 2016. These deployments have covered 19 different areas of expertise, mainly related to resilience, information management, monitoring and evaluation, and communication.

©FAO/S. Glinski

FAO, WFP, UNICEF collaboration

Integrated Rapid Response Mechanism in South Sudan

In early 2018, once again some South Sudanese counties that were in famine in 2017 were hard to access and populations were facing serious food gaps. After months of work to secure safe humanitarian access, a scaled-up response began in July. In order to reach those communities trapped by or seeking refuge from the conflict, FAO partnered with WFP and UNICEF through the Integrated Rapid Response Mechanism to provide immediate and lifesaving support.

Under the Mechanism, FAO delivers emergency response kits, including fast-maturing vegetable and crop seeds and fishing equipment, that are ready to deploy at a moment's notice. Displaced families in remote areas are immediately able to catch fish and rapidly grow food, ensuring food availability and ultimately boosting food security and nutrition. At the same time, partner agencies provided food, safe drinking water, shelter supplies and medicines.

In 2018, FAO reached almost 300 000 extremely vulnerable people under the Rapid Response Mechanism. Simultaneous assessments conducted during the aid operation found that areas were in Catastrophe (IPC Phase 5) before receiving assistance, while those assessed after distributions had better food security conditions, demonstrating that assistance can be the difference between life and death for those living in hard-to-reach areas.

©FAO/A. G. Farran

©FAO/V. Villafranca

Supporting food production

FAO's support to crop and vegetable production for crisis-hit families plays a critical role in restoring livelihoods and rapidly increasing food availability for beneficiary families and their communities.

In the Central African Republic in 2018, FAO provided 1 600 tonnes of crop seeds and 1 tonne of vegetable seeds to more than 250 000 people, who used them to produce just over 20 000 tonnes of cereals and 6 000 tonnes of vegetables. Thanks to FAO's distribution of crop seeds and planting material in Haiti, vulnerable

Scaling up support to vulnerable, conflict-hit families

in Yemen

In 2018, FAO significantly scaled up its assistance in Yemen, reaching over 3 million vulnerable people through a mixture of crop and vegetable seeds, fishing gear, poultry production kits, cash support, animal health campaigns, restocking and animal feed, as well as value chain development. FAO's animal vaccination and treatment campaigns reached over 2.4 million livestock in 2018 despite significant access constraints, providing critical protection from disease and safeguarding livelihoods. In addition, FAO provided training to 104 community animal health workers to provide basic veterinary services in remote areas.

FAO's expertise in farming, livestock, fisheries and forestry is an essential part of the humanitarian response in

Yemen and is not just saving lives but is securing and restoring agricultural livelihoods, which is crucial to address rising food insecurity in the country.

FAO is also focusing its efforts on supporting women, especially given their increasing role in agricultural production since the onset of the crisis. Local-level women's groups are critical partners in reaching the most vulnerable people. FAO is working with them to enhance their capacities to deliver cash, food production inputs and other social services to their existing and new members. By helping to build greater linkages between different women's associations and unions, FAO is supporting greater access to complementary services such as literacy programmes, nutrition training,

social protection and financial service mechanisms. Rural women are thus benefiting from a combination of inputs and training – for example on improved animal health and feeding, milk production and sheep and goat fattening, food processing, backyard gardening and roof water collection – as well as support to strengthen their role in community-level natural resource management, and conflict resolution mechanisms. The promotion of leadership roles for women, for example, in organizing mediation sessions, and advocating for local conflict resolution over water has the potential to reduce not only the violence over accessing water, but enable farmers to make a viable living. In addition, women have become more recognized in their communities as having a role in decision-making processes.

families were able to produce 2 490 tonnes of cereals and pulses as well as 2 800 tonnes of tubers (sweet potato and cassava) in 2018. In Chad, families that received crop and vegetable seeds and planting materials are expected to harvest over 3 500 tonnes of cowpea, groundnut, maize and soybean.

During the 2018 main season, FAO provided around 3.1 million people in South Sudan with 4 800 tonnes of quality crop seeds, which were planted on over 200 000 ha of land, accounting for one-quarter of the total area under cereals cropping in the country. Thanks to FAO's assistance, an estimated 272 901 tonnes of cereal (maize and sorghum) were harvested, accounting for nearly one-third of all cereals produced in South Sudan during last year's main season. Critically, the harvested crops are expected to provide enough cereals for each family for up to six months.

Thanks to the provision of 184 tonnes of rice seeds, together with urea and complete fertilizers, crisis-hit farmers in the Philippines were able to produce more than 2 700 tonnes of milled rice, enough to feed more than 24 000 people in one year. The fertilizers helped farmers to reduce their costs and increase their production. In addition, about 440 ha were planted with the vegetable seeds provided, ensuring an additional source of nutrition and income for vulnerable families.

Supporting main season crop production

in northeastern Nigeria

In time for the main 2018 cropping season, FAO provided agricultural inputs to almost 116 000 families (nearly 900 000 people) in Adamawa, Borno and Yobe States. The support was provided in close collaboration with WFP, which provided food assistance to the FAO-targeted beneficiaries.

Following the season, FAO conducted a post-harvest assessment to determine the impact of the support provided. The vast majority of farmers (93 percent) indicated improved access to land compared with the 2017 rainy season. This can largely be attributed to the improved security situation in areas once dominated by armed groups, although movement restrictions continued to constrain activities in parts of Borno State. For most of those who received support from FAO, farming or livestock rearing was their main source of food and income and for almost two-thirds, FAO was their only source of seeds for planting.

Crop performance was generally described as very good by farming households interviewed. Good rainfall seemingly bolstered crop performance.

As a result of the harvest, families expected to have between four and six months of food, with internally displaced people (IDPs) and female-headed households indicating slightly lower reserves (around three months) as they were more likely to have access to less land. Beneficiary families consumed most of their cereals, selling some pulses and a higher amount of vegetables to generate income.

Protecting livestock

Livestock are a major source of food and income for millions of people across the world, and they can represent a family's entire lifesavings. When disaster strikes and livestock are threatened, it can leave people destitute and without the means to recover. Safeguarding and enhancing livestock assets is a major component of FAO's emergency response, informed and strengthened by broader efforts to build the resilience of pastoral livelihoods. In 2018, for example, FAO's animal health campaigns in crisis contexts reached more than 60 million animals.

In Somalia, where the vast majority of people depend on farming, pastoralism, or a mix of the two as their main source of food, income, and survival, in 2018 FAO safeguarded the assets of over 5.5 million vulnerable people in pastoral communities by vaccinating and treating more than 36 million livestock. Of these, 14 million goats were vaccinated against Contagious Caprine Pleuropneumonia, producing 1.4 million litres of milk per day, enough for 5.6 million people and critical for the nutrition of children and other vulnerable household members. Not only do these large-scale animal health campaigns help people hold on to their assets, they also render livestock more productive, as they provide more meat, milk and increased income for families, communities and markets across the country. Helping pastoralists become self-reliant is 100 times more cost-effective than giving them aid (USD 0.40 per one vaccine dose compared with USD 40 for a new animal).

In the Sudan, FAO's vaccination campaign reached over 6.6 million livestock, protecting them against common diseases, like haemorrhagic septicaemia, PPR, sheep pox and blackquarter. More than 3 300 community animal health workers were simultaneously trained and equipped to provide animal health support to local communities. In Ethiopia, almost 40 000 families benefited from the distribution of animal feed to protect core breeding stock in the face of feed shortages.

In Mali, where conflict and drought have exacerbated already high levels of vulnerability, particularly among pastoral families, FAO provided over 2 000 tonnes of animal feed to safeguard livestock-based livelihoods, alongside training, restocking of herds and an animal health campaign that reached over 600 000 small ruminants.

"Today, I'm not worried anymore, neither of the lean season, nor for the year to come, because by investing in new activities, we will be better able to cope with difficult time"

Sanihan, pastoralist who received cash and livelihoods support from Ségou Region, Mali

In Lesotho, where El Niño-induced drought has exacerbated underlying vulnerabilities, FAO provided 32 tonnes of fodder seeds to 23 grazing associations to help restore pastures for livestock. In addition, over 105 000 livestock benefited from a vaccination and treatment campaign, including vaccinating against anthrax along the anthrax belt in five districts. Thanks to the campaign, no mortality of animals and no new cases of anthrax were reported in 2018, meaning that the outbreak was rapidly contained. In addition, considerable support was provided to enhance the capacity of national extension workers in livestock and rangeland management and disease surveillance.

In Mauritania, almost 1 500 tonnes of livestock feed were distributed to nearly 30 000 people in 2018, together with support to pasture restoration and the vaccination of 180 000 small ruminants. Over 5 000 tonnes of animal feed were provided to vulnerable pastoralists in the Niger in addition to 110 tonnes of wheat bran.

Training on disease surveillance, recognition and reporting was also provided to over 8 000 community workers in Guinea to strengthen the national veterinary health system, together with biosecurity training in veterinary laboratories, evaluation of national animal disease surveillance and implementation of the national brucellosis surveillance plan. In addition, over 100 000 small ruminants were vaccinated. In Sierra Leone, the national veterinary laboratory was equipped, while in Burundi over 700 000 small ruminants were vaccinated.

In Iraq, almost 1 million animals were vaccinated against diseases including PPR, sheep pox, brucellosis, enterotoxaemia, foot-and-mouth disease and lumpy skin disease. Through the campaign, mortality rates were reduced by 95 percent among livestock in Nineveh Governorate that were vaccinated against the diseases.

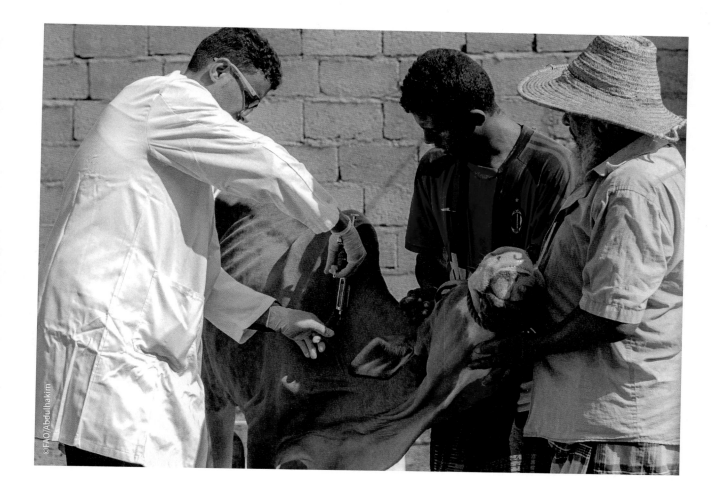

©FAO/Abdulhakim

With FAO's support, the Rapa Livestock Society Group in Kenya produced 5 000 urea-molasses multinutrient blocks for their livestock and surplus for sale. Through this initiative, 50 vulnerable households benefited from subsidized blocks for feeding 250 breeding herds for two months to combat the effects of drought. Overall, the livestock-based drought response in Turkana, Wajir and Isiolo Counties benefited almost 7 300 families (62 percent headed by women) through the provision of animal feed, destocking, meat distribution and collapsible water tanks. In addition, 1.6 million sheep and goats were vaccinated against PPR and sheep and goat pox in Isiolo, Samburu, Marsabit, Turkana, Garissa, Tana River and Kitui Counties.

Tackling PPR and supporting pastoralists
in Eritrea

In 2018, FAO provided support to PPR prevention efforts in Eritrea, helping to strengthen PPR surveillance, detection and laboratory diagnostic capacities across the country. Participatory disease surveillance was conducted with 40 veterinarians and field assistants – the first of its kind in the country. Three national animal health personnel (an epidemiologist and two laboratory technicians) also attended training in South Africa on tissue culture techniques, receipt and preparation of clinical samples, RNA extraction, etc. to enhance their diagnosis capacities. Computers, accessories and reagents were provided to the Ministry of Agriculture to support PPR documentation and reporting, while 30 animal health personnel from six administrative regions were trained and then formulated the National Plan for the Eradication of PPR by 2030. In addition, a national PPR communication strategy was developed, alongside relevant materials (booklets, posters, leaflets), as well as video spots on PPR in two local languages that were broadcast on local TV to raise awareness of the disease.

In addition, FAO assisted over 100 000 drought-hit pastoralists in Gash Barka region to protect their livestock by vaccinating 165 000 cattle against lumpy skin disease and Brucella and some 400 000 sheep and goats against PPR and sheep pox.

©FAO/Dominica

Restoring and diversifying livelihoods

FAO also restores agriculture-based livelihoods in the aftermath of natural disasters, including Hurricanes Maria and Irma, which hit the Caribbean in September 2017. In Dominica, damages and losses to the agriculture sector as a result of Hurricane Maria were estimated at USD 211 million. In 2018, FAO assisted 14 000 hurricane-affected families to restore their crop, livestock and fisheries production. At the same time, the Organization supported national capacity building for data gathering, analysis, evaluation and reporting on damage and loss caused by disasters in the agriculture sector, which is crucial for timely and informed decision-making and disaster risk reduction efforts.

In Cuba, where over 50 000 ha of crops were lost to Hurricane Irma, as well as major damage to the livestock sector, FAO helped to reboot local food production, reaching more than 1 million hurricane-affected people in 2018.

The FAO programme focused on restoring short-cycle food production – for example rehabilitating 260 greenhouses, as well as 51 poultry and pig production facilities – ensuring the availability of eggs, pork and vegetables. The impact of FAO's activities was amplified owing to the close alignment of interventions with the Government's response programme.

In Myanmar, FAO helped to increase the nutrition and dietary intake of 14 424 flood-affected families in three regions/states by providing support to vegetable production.

In 2018, continued large-scale population movements driven by conflict, food insecurity and economic shocks had major implications for the food security and livelihoods of both displaced and host communities. In Cox's Bazar, Bangladesh, FAO was one of the first actors to focus on supporting host communities as well as refugees – a critical step in mitigating tensions and supporting social cohesion.

In Kenya, recognizing the impact of refugee populations on local natural resources, FAO provided ten new charcoal producer groups with one steel kiln each and members were trained on sustainable charcoal production and business management. The groups were introduced and linked to Kakuma and Kalobeiyei refugee markets, while 2 000 refugee households (82 percent female headed) received fuel-efficient stoves and were trained on their use. In Uganda, over 7 000 refugee and host community households received fuel-efficient stoves.

With a massive movement of Rohingya into Cox's Bazar, there has been a serious concern about the potential for rising tensions between the refugees and host communities caused by increased demand on already scarce natural resources, food and employment. FAO is supporting the Government of Bangladesh to promote social cohesion and improve the food security of both Rohingya refugees and host community populations, through a coordinated, targeted and equitable response. In 2018, FAO provided assistance to more than 87 800 refugee and host community households (395 000 people).

This support has included crop and vegetable production assistance to ensure host communities can produce nutritious food for themselves and Rohingya families while earning extra income, including by setting up 60 farmers' groups and linking them to new refugee markets. The improved vegetable production of 1 500 farmers has increased availability in the local market, thereby enhancing the nutritional intake of the local population. Sales of the produce enabled many farmers to invest in better equipment, purchase natural fertilizers and high quality seeds. Farmers have also received training in climate-sensitive production techniques (greenhouses, vermicomposting) and drip irrigation systems have been introduced. Micro-gardening has also been promoted to enhance nutrition and make use of limited land areas by providing seeds and tools to among 25 000 refugee families and 25 500 host community families. In addition, over 60 000 families from both communities have received storage drums to protect their food from infestations and prevent food-borne illnesses. FAO has

©FAO/K. Coa

also been working with the national Forest Department to mitigate the risk of landslides and set up plant nurseries to replant degraded and deforested land due to the increased demand for firewood for cooking. This means a reduction in the risk of landslides and flash flooding, especially during heavy rains and monsoon seasons. FAO is also creating opportunities for alternative livelihoods for host communities in the fisheries sector, promoting sustainable aquaculture production through capacity building and inputs for fish pond cultivation.

Building skills of refugees and host communities
in Turkey

In Turkey, FAO's programme focuses strongly on developing skills of both Syrian refugees and host community members through vocational trainings and facilitating employment opportunities in agriculture.
The training in food processing and agricultural production techniques combined with efforts to link trainees to jobs is well recognized by the participants, local government and

partners like UNHCR. The training was developed following an assessment of labour market gaps in some areas with high concentrations of Syrian refugees and, in 2018, included vegetable and crop production, livestock and poultry management, beekeeping and food processing practices and reached almost 1 500 families. The trainees are then invited to participate in job fairs and linked to local companies

seeking their specific skills. Thanks to the training, to-date about one-fifth of participants have found employment, while over 80 percent were happy with the training and the skills provided, which are meeting a critical gap in local labour markets. However, this is just the start and in 2019, FAO plans to continue to upscale this programme in collaboration with the Government and other partners.

In response to rising food insecurity among families hosting migrants from Venezuela in Colombia's La Guajira region in 2018, FAO assisted 7 000 vulnerable, indigenous Wayúu people who were often unregistered and therefore not receiving any other form of assistance. In agreement with local communities in the rural municipalities of Manaure, Maicao, Albania and Riohacha, FAO provided agricultural inputs (veterinary supplies, seeds, tools, irrigation systems), rehabilitated water supply systems for human consumption and agricultural use, and supported animal health campaigns. In less than six months, the programme helped to rapidly restore and enhance food production, laying the foundations for social, cultural and economic integration of communities from both sides of the border.

FAO has also been supporting alternative livelihoods for crisis-hit families to help them produce more food and better cope with the next shock; for example, promoting fish farming integrated with rice farming in Mali by installing floating cages, and providing fish farming kits to vulnerable families in northeastern Nigeria's Borno State, together with training.

More on the promotion of fish farming in northeastern Nigeria

©FAO/P. Wiggers

In the Republic of the Congo, FAO supported the rehabilitation of the state fish farm for the reproduction of fry, which will lead to the production of 200 tonnes of tilapia, a very important source of protein for vulnerable communities. The fry will be provided to 470 small-scale fish farmers – 60 percent of whom are women – to enhance their production. More than 6 000 people in the Sudan also benefited from fishing kits in 2018.

Following a disaster, household savings, productive capacity and assets are often lost, meaning people can no longer meet their basic needs. In these crisis situations, cash assistance is critical to address immediate food needs and mitigate displacement. In Somalia, FAO's cash programmes, including the 24-week cash-for-work programme and the three-month cash+ programmes (unconditional cash coupled with agricultural/pastoral/fisheries livelihood-support packages) reached 154 252 households throughout 2018.

Through a cash-for-work programme in Haiti, about 40 000 people had cash in their pocket and 31 community nurseries produced 2.3 million of fruit and agroforestry seedlings. More than USD 670 000 was disbursed through cash-for-work schemes in Iraq that cleared over 40 km of water pathways and 360 ha of agricultural land, cleaned 5 km of drainage canals and 102 km of irrigation canals, and set up 1 ha of home gardens. Almost three-quarters of the cash-for-work participants reported they are already making use of the repaired and rehabilitated infrastructure, especially water from the irrigation scheme and vegetable production on the established home gardens. They also reported and approximate increase in household income of 135 percent. For the majority – 80 percent – the cash-for-work scheme was their main source of income during the implementation period, with 65 percent reporting that the cash received was adequate to meet their basic needs, where 45 percent of the amount received was spent on buying food and the rest spent on health and debt reimbursement.

"The income enabled me to pay medical expenses for my parents, to buy food and afford my family's basic needs."

Adela Thaweb, cash-for-work participant from Edriskhabz village, Iraq

In Burkina Faso, a total of USD 1.8 million was provided to 80 000 people through unconditional cash transfers, while a further USD 148 000 was disbursed under a cash-for-work programme that reached 2 500 households. In the Niger, almost USD 80 000 was provided as unconditional cash to support pastoral families in crisis, while a cash-for-work programme rehabilitated community assets, including a 1 200-km firewall.

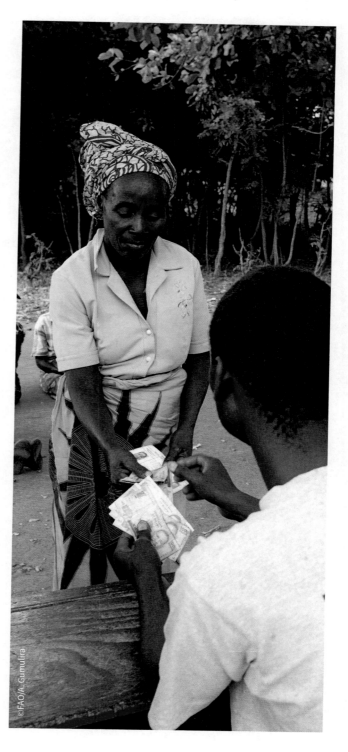
©FAO/A. Gumulira

Coordinating the food security response in emergencies

At the start of 2018, more than 110 million people in emergency situations were severely food insecure. Of these, 74 million people were targeted through Humanitarian Response Plans. In the global Food Security Cluster's main operations – in Nigeria, Somalia, South Sudan, the Syrian Arab Republic and Yemen – partners jointly supported around 21.5 million people through food assistance and more than 15.5 million through livelihoods assistance. Almost one-third were reached through cash and voucher programmes.

Global Food Security Cluster activity in 2018

Food security coordination structures

To save lives and livelihoods and to ensure the dignity of people affected by humanitarian crises, Food Security Clusters/Sectors coordinate the food security response in emergencies.

The global Food Security Cluster is co-led by FAO and WFP. It supports some 30 in-country Food Security Clusters/Sectors. Country-level Food Security Clusters/Sectors coordinate the food security response during a humanitarian crisis by ensuring that the Cluster Lead Agencies, FAO and WFP, other United Nations agencies and international organizations, NGOs, civil society, resource partners and government representatives work hand in hand to reach people in need. The objective is to support food security partners in reducing the number of people in humanitarian need with food assistance and resilience activities.

In 2018, the global Food Security Cluster convened more than 1 200 partners at country level, among which more than half are national partners.

Coordination of the food security response is cost-effective – it is achieved using just 30 cents per USD 100 of resource partner funding to support its country-level coordination teams.

In 2018, the global Food Security Cluster scaled up the training (in English and French) and deployment of new cluster coordinators and information management officers to respond rapidly to the country lead agencies' request for support. Current cluster coordinators and information managers are also coached with advanced training through retreats organized at regional level.

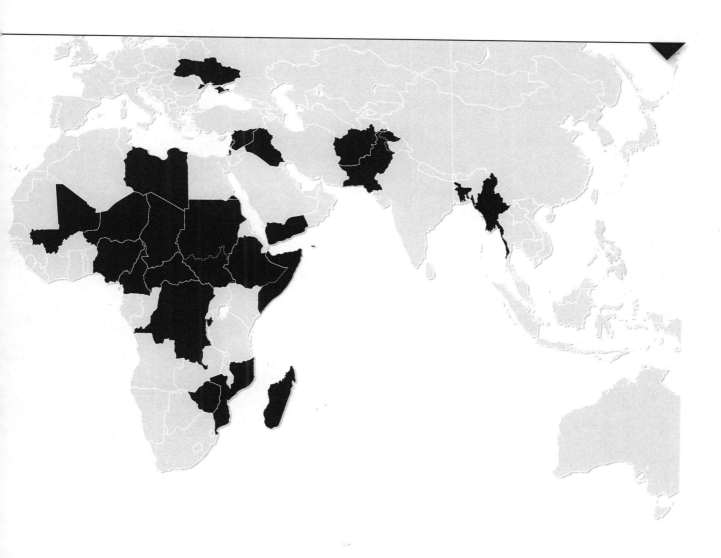

Safe Access to Fuel and Energy (SAFE)

Crisis-affected populations – including refugees, IDPs and host communities – often have severely constrained access to fuel and energy for cooking, heating, lighting and powering. FAO is working with partners through the SAFE initiative to address energy needs during emergencies and protracted crises, and to build resilient livelihoods in a sustainable manner. The SAFE initiative is coordinated through the SAFE Humanitarian Working Group, of which FAO is an active member. FAO's work on SAFE has gained momentum in recent years, due in part to the increased recognition of the importance of energy issues for crisis-affected populations, both in the field and in major global policy processes, such as the Global Plan of Action for Sustainable Energy Solutions in Situations of Displacement.

Globally, nearly 3 billion people rely on traditional biomass, such as fuelwood, charcoal, animal waste and crop residues as sources of fuel for cooking and heating. Refugee and IDP camps are often established on fragile, sparsely forested land, and both host and displaced populations depend on the scarce natural resources found in areas surrounding the camps. The influx of displaced people, leading to increased cooking fuel needs, can often exert great pressure on forests and woodlands.

This is frequently a source of tension between host and displaced communities, which increases the risk of women and children being harassed and assaulted while searching for fuelwood. Furthermore, collecting fuelwood takes time away from school attendance, income-generating activities, childcare and leisure, which can reduce the effectiveness of other humanitarian and development programmes targeting women and children. The lack of sufficient cooking fuel also has an impact on the nutrition and health of vulnerable households as women may resort to undercooking food or skipping meals to save fuel as well as bartering food for fuel. In displacement contexts, food is often cooked on a three-stone fire in poorly ventilated spaces, which exposes women and children to respiratory illnesses. Both host and refugee populations rely on short-term and unsustainable livelihood activities such as charcoal production and selling of fuelwood.

©FAO/L. Beshara

In 2018, FAO and its partners have assisted more than 200 000 people by supporting clean cooking through different forms: fuel-efficient stoves reducing charcoal and wood consumption, construction of household biogas digesters and the establishment of a Liquefied Petroleum Gas (LPG) value chain to reduce pressure on forest ecosystems. FAO has also supported solar irrigation to support livelihood as well as solar portable lighting to improve protection when collecting woodfuel. In addition, FAO has established forest management plans, for example around refugee camps in Ethiopia, Kenya and Uganda, which take into account the energy demands of both displaced and host populations.

In 2018, FAO supported improved energy access through the SAFE initiative in Bangladesh, Chad, Djibouti, Ethiopia, Jordan, Kenya, Nigeria, Palestine, Somalia, South Sudan, the Sudan, the Syrian Arab Republic, Uganda and Yemen.

FAO's support to improve energy access in 2018

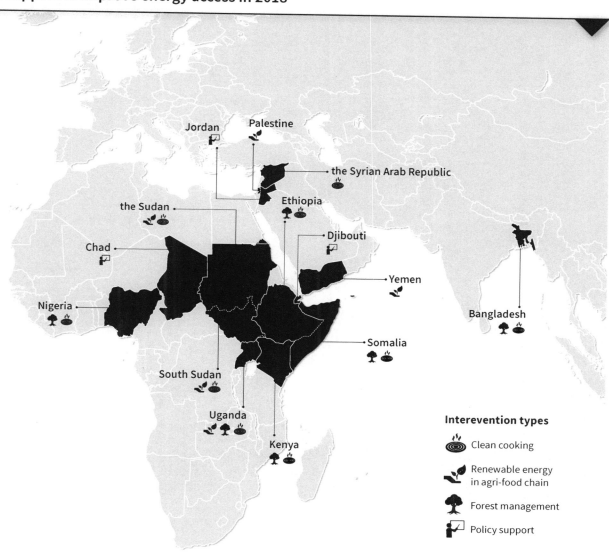

Interevention types

Clean cooking

Renewable energy in agri-food chain

Forest management

Policy support

Addressing reforestation needs in **Bangladesh**

FAO assisted 2 500 refugee households from the Kutupalong settlement and 500 host families located in the adjacent areas to the camps. The beneficiaries received a cooking stove, a full LPG cylinder and refills for five months. Communities also received training in efficient cooking training, while nutrition awareness-raising activities were conducted. A fire warden system was set up following a training on fire safety and use of LPGs. In addition, support to environmental restoration and reforestation for land stabilization was

carried, including distribution of tree saplings, seedling production, grass and shrub cultivation for land stabilization, as well as monitoring the reforestation activities.

FAO's activities are implemented within the framework of the SAFE Plus project, in partnership with IOM, WFP and UNDP. This three-year multi-sectoral, multi-partner project addresses cooking fuel needs, environmental restoration and improves food security and livelihoods.

©FAO/GMB Akash

Supporting conflict-hit families in **northeastern Nigeria**

FAO established three fuel-efficient stove production centres, employing 100 people to manufacture 18 000 fuel-efficient stoves to be provided to vulnerable households in stable and accessible areas. By the end of 2018, more than 6 000 stoves had been distributed. FAO also provided 2 500 SAFE kits comprising portable, durable and lightweight stoves and solar lanterns for highly mobile IDP and returnee households in unsecure areas. The provision of solar lanterns boosts protection, particularly among IDPs. FAO is also supporting the rehabilitation of tree

nurseries and solar irrigation among displaced and host communities. Cooking using the traditional stove can be cumbersome. According to Mallama Abubakar, a resident of Bakassi camp, it can even be dangerous:

"We are happy to get rid of the Murfu," she said, referring to the Hausa (local language) name for cooking on three or more large stones propping up heavy pieces of firewood. *"It is making many of us women sick, especially in the chest and eyes,"* she said, motioning emphatically to her lungs.

Helping meet energy needs in the **Syrian Arab Republic**

Due to the ongoing crisis, the high dependence on woodfuel in the Syrian Arab Republic, especially in rural areas, is resulting in an alarming rate of deforestation. The crisis has also increased energy prices and the burden they represent on families' income. For this reason, FAO has supported the construction of 60 household biogas digesters for farms with at least three to five cows, supporting 500 people. The digesters yield enough to cover the needs of a normal Syrian family for cooking and heating, as well as providing other benefits such as the digestate, which is used for fertilization. The digesters have also helped to reduce pests associated with manure and odours.

"We used to throw our cows' waste next to our house. The smell was awful. This intervention contributed to reduce the smell. I now have gas for cooking and heating for my family and my son's house too. We also produce organic fertilizers and we use them for our lemon and olive trees".

Hassan Mahmoud, local farmer

©FAO/The Syrian Arab Republic

Improving domestic water uses in **Yemen**

Acute fuel shortages in Yemen have a severe impact on household livelihoods, particularly on health, transport and agricultural production. The absence of fuel to power shallow well pumps could have important health impacts as communities may resort to irrigation with untreated wastewater, which can increase the risk of water-borne diseases such as cholera.

FAO has supported the installation of solar photovoltaic pumps to ensure continuous water provision to rural households for domestic use, support to crop and fodder production and livestock watering. The systems ensure water provision for farming households (irrigation and drinking) and overcome the lack and high costs of fuel.

To ensure sustainability, FAO has supported the replacement of fossil fuel pumps with solar pumps in existing wells, the establishment or reactivation of Water Users' Associations, facilitated

the establishment of water-sharing agreements and installed water meters and systems without batteries, as well as efficient irrigation systems such as drip irrigation. In areas affected by irrigation from untreated wastewater, the Water Users' Association monitors the community ban on using untreated wastewater for irrigation and conveys public health messaging. This is particularly relevant given the recent cholera epidemic in the country.

"We fear for our children from water-borne diseases, especially cholera, but we had no other choice to make a living. This land is all we have and agriculture our only source of livelihood. We were confronted to a difficult choice: use sewage water or leave the land without irrigation as we cannot afford to pay the cost of fuel."

Abdullah Handal, local farmer

Cash to restore food production and access

Cash and vouchers play a critical role in FAO's response to shocks and crises, when farmers, pastoralists and fishers can no longer buy food or the productive inputs they need because their assets have been damaged or depleted. FAO's cash-based transfers provide immediate relief to farmers, and contribute to strengthen the resilience of their livelihoods to future shocks (e.g. drought, poor production, etc.), increasing agricultural production, improving food security and nutrition, and reducing rural poverty. They support the transition from humanitarian assistance to development, including through enhanced linkages with social protection systems that can be leveraged to respond to shocks and crises.

For the last 15 years, FAO has been using cash-based transfers in almost 50 countries, using modalities such as input trade fairs, voucher schemes, cash-for-work, unconditional and conditional cash transfers and since 2013, cash+. In 2018, FAO delivered cash-based transfers using these modalities in 26 countries through 66 operational projects, reaching over 2.2 million people.

Countries such as Nigeria, Somalia or South Sudan have extensively used cash-based transfers to assist vulnerable, crisis-affected populations.

In northeastern Nigeria, the Boko Haram insurgency has led to heightened levels of displacement and food insecurity. While humanitarian access is improving, most displaced families still rely on vulnerable host communities for basic needs, including food. This has put already impoverished host communities under extreme pressure, leading to increased exposure to food insecurity and malnutrition. Within this context, among other activities, FAO provided 10 000 households with cash transfers (USD 100) to purchase essential food items and meet their basic food needs in Adamawa, Borno and Yobe States. FAO reached 15 920 households (106 400 people) with this modality in 2018 country-wide.

In Somalia, the 2016/17 drought devastated production, productive assets and income sources. Low production also left farmers without seeds to plant, cut wage labour rates by at least 50 percent, and increased food and water prices. This was aggravated by soaring prices of staple cereals, brought on by consecutive seasons of poor production, which continues to restrict normal food access. Modalities such as cash-for-work and unconditional cash transfers allow families to prioritize meeting their immediate needs while restoring their food production. Cash-based

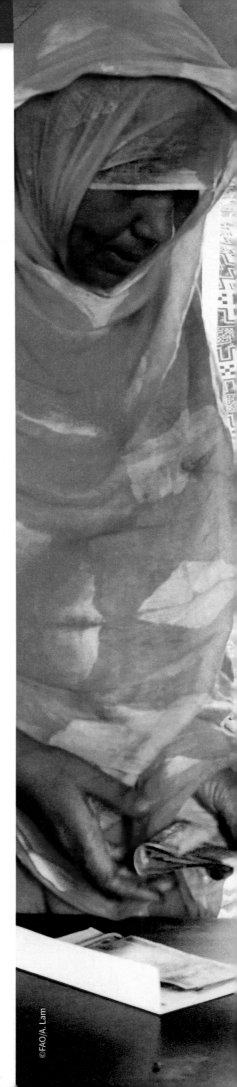

©FAO/A. Lam

interventions were especially vital in rural areas, where families lacked food and income to buy food due to extensive crop failure and livestock losses. During 2018, FAO delivered cash-based transfers worth USD 26.7 million reaching 154 252 households throughout the country.

In South Sudan in 2018, soaring food insecurity and an elevated risk of famine required a targeted livelihoods response to support local food production. The organization of seed fairs were prioritized when targeting the households because they enable farmers to purchase their preferred seed variety from local seed producers, based on seed availability and security conditions. In 2018, 77 311 households (463 866 people) received seed support through this mechanism.

FAO cash-based interventions in 2018

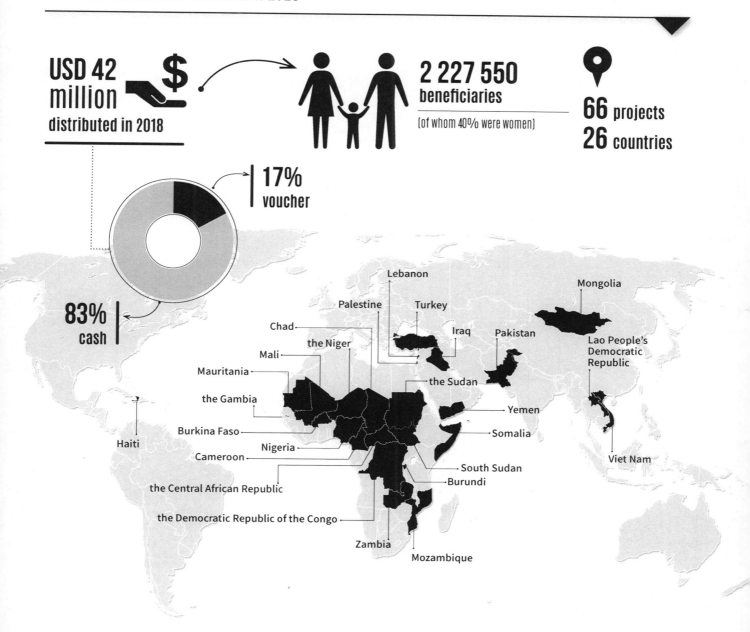

USD 42 million distributed in 2018

2 227 550 beneficiaries
(of whom 40% were women)

66 projects
26 countries

17% voucher

83% cash

Lebanon
Mongolia
Palestine
Turkey
Chad
Iraq
Pakistan
the Niger
Lao People's Democratic Republic
Mali
Mauritania
the Sudan
the Gambia
Yemen
Burkina Faso
Somalia
Haiti
Nigeria
Cameroon
South Sudan
Burundi
the Central African Republic
Viet Nam
the Democratic Republic of the Congo
Zambia
Mozambique

❝ Voices of **Malian** pastoralists

"Our food reserves started depleting and we had to sell three sheep... FAO's intervention has come at the right time because the cash allowed me to replenish my little breeding. I received XOF 90 000 (USD 153) which allowed me to buy three goats and animal feed. This allowed me to establish my herd... As well as the cash, I was also trained in breeding. Now I know how to take care of my animals."
Nana Diawara

"I received XOF 90 000 (USD 153). I bought a cattle and animal feed. I am going to fatten the cattle and sell it. I am planning to make several fattening cycles. I do not want to solely depend on agriculture. FAO's support allowed me to diversify my activies. Today, I am not afraid to lack money anymore."
Assia Diallo

©FAO/S. Nguyen

Cash transfers in **northeastern Nigeria**

In northeastern Nigeria, the cash provided by FAO was used by 73 percent of recipients to buy food, while others used the money to buy clothes for family members, invest in livestock and in small businesses.

Overwhelmingly, cash transfers enabled vulnerable families to store rather than selling their harvests, and also meant they could spend their time working on their own farm rather than doing casual labour on other farms. Thanks to the cash provided, 83 percent of the beneficiaries stated that they consumed at least three meals a day, a striking improvement from the only 50 percent that were consuming three meals per day prior to the intervention.

Lami Abakar, a mother of four, lost her husband due to the insurgency in Banki village, in Nigeria's Borno State. She then moved to Dallori camp and has struggled to meet her families' needs, begging for food that was barely enough for two meals a day.

The assistance provided under FAO's fresh food voucher scheme was the first assistance she received. The fresh food vouchers essentially provide families with vouchers to purchase fresh food in local markets, which improves their families' diets and nutrition. At the same time, participants were also provided with a small cash transfer to meet other needs.

Cash-for-work in **Iraq**

©FAO/C. Yar

Farmers in Iraq's Anbar Governorate depend hugely on a canal that provides irrigation water to more than 20 villages. Conflict and displacement have significantly damaged public infrastructure, including canal systems, and the *"resources with the Government alone are not sufficient to re-build infrastructures in these areas having countless damages,"* according to Senior Engineer Mudher Abed.

FAO managed a public works programme in Anbar Governorate, providing cash transfers to vulnerable, food-insecure and/or crisis-affected households in return for labour. Working in close coordination with the Ministry of Agriculture, FAO identified 320 vulnerable families in Saqlawiyah village to participate in the cash-for-work scheme. Under the programme, 23 km of canals were rehabilitated, providing irrigation water to more than ten villages, while 108 home gardens were established for female-headed households and 0.53 ha of land developed and prepared to grow seasonal vegetables. *"Due to destruction of this canal, the farthest farmers were not able to get water, which resulted in uncultivated lands. By cleaning the canal and its tributaries, the water now reaches the agricultural land and farmers started to expand cultivation land and will grow crops this season. Up to 10 000 acres of agricultural lands in Saqlawiya sub-district will benefit from a better irrigation system,"* said Mr Abed.

In addition to their seven children, Ami Mohammed and her husband take care of 22 orphans in Bakassi camp, after fleeing Marte, Borno state. Meeting the needs of the children has been challenging for Ami and her husband, who lost his livelihood since the insurgency and often has to borrow money.

They were very happy to have benefited from FAO's fresh food voucher scheme which for the first time allowed them to receive both food and money at the same time, twice. They were able to purchase more (maize, rice, oil, etc.) thanks to the cash received, which was also used to pay off debts.

©FAO/Nigeria

As part of the cash-for-work programme, some of the most vulnerable households identified were targeted with unconditional cash transfers.

©FAO/N. McNally

Cash modalities in **Somalia**

Cash-for-work and unconditional cash transfers

Total households reached

67 264

45 210 households reached through **cash-for-work** ···············USD **11.4 million**

9 907 households reached with **unconditional cash transfers** as part of the cash-for-work ···············USD **2.8 million**

12 147 households reached for **pure unconditional cash transfers** ···············USD **3.4 million**

Total cash distributed

USD 17.6 million

Total infrastructure rehabilitated

325

≈ **18** canals

74 contour bounds

233 water catchments

Cash+

cash + fish

1 000 households reached with

 USD 314 028 in cash assistance

 1 000 fishing kits

51 solar fridges (50 solar fridges to beneficiaires, 1 to the ministry).

 80 women from urban and peri-urban areas trained in fish handling, preparation and preservation

cash + crop

79 099 households reached with

 USD 7 885 028 in cash assistance

 Agricultural inputs 564 tonnes of maize, 536 tonnes of sorghum, 634 tonnes of cowpea, 12 tonnes of vegetable seeds, 2 928 tonnes of fertilizer, 636 000 storage bags, 3 250 ha prepared with tractor hours and 7 050 ha irrigated

cash + livestock

6 889 households reached with

 USD 853 616 in cash assistance

 6 889 livestock inputs

Reducing risks and addressing vulnerabilities

Disasters and crises do not just have immediate, short-term effects on people's lives and livelihoods. They can destroy livelihoods that have taken generations to build. Decades of marginalization and environmental degradation, exacerbated by recurrent shocks driven by climate change and conflict undermine people's ability to cope with the next stressor, for example turning a dry spell into a major humanitarian crisis.

That is why FAO is stressing proactive risk management and seeking to address the underlying fragilities that leave millions of people so exposed to the effects of shocks and stresses.

FAO's resilience work encompasses several different and complementary interventions including disaster risk management and disaster risk reduction, and goes further by including conflict-sensitive approaches and social protection measures as well as financial risk transfers, land and natural resources tenure/access, climate change adaptation to extreme events, etc.

The evidence shows that when farmers apply disaster risk reduction practices, like conservation agriculture or the use of drought-tolerant seeds, they see returns 2.5 times higher than for their usual production practices. Scaling up vulnerability reduction measures is crucial to strengthen agricultural livelihoods and reduce the impact of shocks, so that people can protect and rebuild their livelihoods better and faster, contributing to lasting food security and nutrition.

This means investing in climate-resilient agriculture and food systems – from enhancing risk monitoring and early warning systems to promoting and facilitating the adoption of climate-smart production techniques, strengthening risk-sensitive and shock-responsive social protection schemes, enhancing national and local level policies and practices, and building risk management into emergency preparedness and response (e.g. strategic reserves of seeds, food and animal feed) so families can recover faster and better.

In Somalia, after two years of poor harvests and massive livestock losses following drought, exacerbated by floods in early 2018, recovery will take time for affected families. While the humanitarian response to avert a famine was critical, farmers and herders need continued support to emerge from debt cycles, restore their livelihoods and be better prepared to cope with the next shock. This requires investing in building more resilient, diversified livelihoods. A farmer who relies solely on producing one crop can lose everything to a flood or a pest outbreak, like FAW. FAO has therefore been promoting fishing among farming communities who live along rivers and supporting beekeeping for vulnerable women. Honey can generate a good income and beekeeping requires relatively little investment of time, which is important for women who already dedicate most of their time to household duties and childcare.

Recognizing the importance of providing young people with genuinely viable livelihoods, FAO has also been engaging them in new opportunities, such as management of Prosopis – an invasive species that can overrun cropping areas and grazing lands but that can also meet high demand for fuelwood and charcoal. While previous efforts have been made to make use of the extremely sturdy plant, most communities in Somalia struggle to manage its spread, which can significantly hamper local livelihoods. Under FAO's programme in 2018, some 2 480 youth received USD 595 400, along with training and equipment, to manage an area of 199 752 ha of Prosopis. As a result, the spread of the plant is being better contained, while young people are able to generate an income from the sale of Prosopis for fuelwood, building poles, fences, and other purposes.

Building resilience and sustainability
in Lower and Middle Shabelle, Somalia

Building resilience is not only about safeguarding livelihoods: it is also about giving farmers, pastoralists and fishers more options to diversify their economic and productive activities, reach new markets and add value to their production. Preventing the degradation of natural resources and restoring the ecosystems that underpin agriculture, fisheries and pastoralism are crucial in achieving this and ensuring vulnerable rural people not only have the means to support themselves but to prosper.

Just two years ago, Somalia was on the brink of famine during one of the worst droughts the region has witnessed. The severity of the 2016/17 drought – plus unexpected flood and cyclone events in May 2018 and a return to drought conditions by yearend – underscored the importance of addressing water management in Somalia. Key resource partners responded with multi-year investments in late 2018 to restore irrigation and support agriculture value chains in Lower and Middle Shabelle regions of Somalia's southern breadbasket. These efforts, although

just started, will ultimately better connect smallholders to the resources, services and networks they need for a more a resilient and prosperous livelihood – including improved access to water, inputs, extension services and markets.

The programme will rehabilitate a total of 400 km of major and minor canals managed by newly established or strengthened water management committees. Beyond restoring irrigation to 25 000 ha of farmland, these efforts will also be transformational in building community resilience against recurrent climate shocks. The repair and effective management of irrigation networks will reduce farmers' vulnerability to drought conditions, while the removal of silt deposit from the canal basins will mitigate against river overflows and related flood events.

Most farmers in the project areas have lacked reliable access to irrigation water for decades. This has made them more vulnerable to drought and dependent on rainfed production –

diminishing the type and volume of crops they can produce. However, even with restored irrigation, individual smallholder farmers produce primarily for subsistence. Alone, they are invisible to extension, supply chain and market actors. FAO will therefore support these farmers to form strong producer groups. By aggregating needs and production, farmer groups can achieve the economies of scale needed to access better input supply chains and link to more profitable markets.

This is not a standalone, isolated programme: it contributes to the Ministry of Agriculture and Irrigation's Strategic Plan 2016-2020 and builds on repair work already being carried out by other partners in the region. These investments will serve as a foundation and catalyst for other actions in the region, to establish more robust and resilient livelihoods through sustainable access to water and aggregated production, addressing specific bottlenecks in the agriculture value chain that impede development.

Water scarcity and poor water management practices are a major factor influencing local-level instability and rising hunger in many contexts. In Afghanistan, rural populations are on the forefront of climate change, as well as being highly exposed to the impact of violence and displacement. In response to the worst drought that Afghanistan has faced in a decade, the FAO-led livelihoods response reached 500 000 acutely food insecure people with timely agricultural and livestock support.

This enabled farmers to cultivate in the main planting season and herders to protect their animals ahead of the hard winter. This means building on years of experience in the country and adapting best practices from development to the most vulnerable contexts to address the root causes of this vulnerability.

Adapting to climate change through improved water and natural resource management is key to more resilient and sustainable food systems in Afghanistan. For example, for years FAO has worked extensively on rehabilitating Afghanistan's national irrigation system and strengthening institutional capacity to better manage water resources; training more than 7 000 staff from national institutions on advanced water management issues and providing irrigation to almost 1 million additional ha of farm land and setting up hundreds of local committees to manage water use and irrigation schemes.

One million cisterns
in the Sahel

Efficient and sustainable management of water resources is a priority to improve the resilience of vulnerable communities. The "1 million cisterns for the Sahel" initiative promotes simple and cost-effective rainwater harvesting and storage systems for vulnerable communities, especially women.

In 2018, FAO together with partners and communities, launched the pilots of the programme in the Niger and Senegal.

In Senegal, 16 cisterns were constructed for families (each measuring 15 m³) and three cisterns were constructed for agricultural production (each measuring 50 m³) at the community level, reaching altogether 360 beneficiaries. In the Niger, FAO constructed five family cisterns and six community cisterns for agriculture reaching altogether 500 beneficiaries. Family cisterns cover household drinking water needs during the entire dry season, and can also be used to nourish micro-gardens. Community cisterns, on the other hand, cover water needs for agricultural production during recurring dry spells during the rainy season and also ensure an additional cycle of production for 0.5 ha of land.

In addition to access to water, women benefit from climate-resilient inputs for vegetable gardening and training, especially using and maintaining cisterns. The project also contributes to providing safety nets as local communities receive cash in exchange for their labour for constructing the cisterns.

In the coming three years, the project is expected to reach an additional 10 000 women in Senegal, 5 000 in the Niger, and another 5 000 in Burkina Faso.

"We started farming during the dry season barely two years ago. Now we can also produce vegetables for sale during the dry season, including salad, onion, chilli pepper, eggplant, mint and okra."

Like many farmers in Senegal, Guilé Mané used to struggle through the dry season. Rainfall here can be very low and irregular, even in the rainy season. "We were working in the fields during the rainy season, but we did not do anything during the dry season," says Guilé, 39, who heads a farmers' association called Diapo Ande Liggeye (United to Work) in her home area of Keur Bara Tambédou. The lack of water meant crop and food shortages, more frequent illnesses, and insufficient income from the sale of whatever the

farmers managed to grow. "Sometimes parents could not even pay their children's monthly school fees." Guilé and the other women had to walk long distances to reach sources of clean water and use part of their income to pay for it.

The women, their families and local masons were trained to build cisterns for year-round water storage. The cisterns hold water harvested from a collection area such as the rooftop of a hangar or shed. The beneficiaries and masons were paid for their work while the farmers received training in climate-smart agricultural practices.

Guilé says she and the other farmers have learned several new things from FAO and its partners, such as how to draw up a gardening plan, how to set up a nursery, and how to keep the soil and plants healthy. "Now we can do everything by ourselves."

The programme has also empowered women financially. Through the farmers' association, the women have created a fund with proceeds from their market sales. Each woman is able to withdraw money to meet household or personal expenses before paying it back at the end of the month.

©FAO/S. Glinski

In the Syrian Arab Republic, over 25 500 families have either remained in their villages to resume food production or returned to their lands in 2018 thanks to FAO's irrigation and water resource rehabilitation projects in areas of Aleppo, Hama, Homs and Tartous Governorates. The rehabilitation of damaged gates, control intakes, water canals, pumping stations and installation of generators, has transformed about 20 000 ha of lands from low production to fruitful green lands.

Adopting climate-smart agricultural practices, like the use of drought-tolerant seeds, further spreading conservation agriculture techniques which have been shown to not only use less resources but also to significantly increase productivity, increasing water storage facilities, improving flood protection of irrigation systems, harvesting flood and rainwater, managing at the watershed level, reforesting lands to reduce soil erosion and recharge groundwater, developing climate resilient policies, establishing early warning and awareness systems, and building the capacity of farmers on these will all play a vital role in achieving resilient livelihoods and averting future food crises.

In southern Africa and in Indonesia's Nusa Tenggara Timur and Nusa Tenggara Barat Provinces, FAO has been implementing multi-year programmes to strengthen livelihoods' resilience against climate-related threats and crises by promoting climate-smart agriculture, specifically conservation agriculture, and increasing understanding of disaster risk reduction. This has involved a combination of providing seeds, fertilizers and farming equipment designed for conservation agriculture (such as jab planters) together with intensive training through farmer field schools or a lead farmer model. In Indonesia, where farmer field schools have been set up in 28 districts, promotion of conservation agriculture techniques has had a significant impact on yields, which increased from 4 tonnes/ha in 2017 to 5.9 tonnes/ha in 2018, on average.

Strengthening the resilience of livelihoods
in South Sudan

Despite the ongoing conflict in South Sudan and massive humanitarian needs, there remain considerable opportunities to undertake longer-term support for resilient livelihoods and food systems. FAO's programme in the country thus incorporates activities like support to local seed production and seed fairs and nutrition to link local producers to markets, as well as huge investment in the national veterinary cold chain and collaboration with UNESCO to develop and implement an education programme specifically designed for pastoralists.

One example of this work in 2018 was the support to the farmer field school for the Illaya community in Torit. Community members received a solar pump, along with training on its operation and maintenance, alongside hands-on training on seedling and vegetable production. At the same time, the group members received agricultural tools as well as seeds. Following the training, the participants were able to plant vegetables on 2 acres and are using the pumps to irrigate the vegetable fields.

For the Murwari community, which relies on declining subsistence-level production to meet their food and income needs, FAO provided a range of support to enhance their resilience. A multi-purpose solar-powered deep borehole was constructed, which provides clean and safe water for domestic consumption, crop irrigation and livestock production. In addition, the villagers were trained in improved production techniques through agropastoral field schools and linked to extension services, who provide regular follow-up assistance and advice. The field schools set up village savings and loans associations that have helped members to start new income-generating activities, as well as starting up a community nursery for tree and vegetable seedling production.

With FAO's support, the community has seen significant changes. Members have increased land for vegetable and crop production and gather fresh produce from their own farms. It has also been easier for each member to preserve a good quantity of seeds for the next year's planting, while farmers have sold surplus produce at the market, which has encouraged them to start on-farm sales of produce, to purchase the products they need.

Even in complex contexts, FAO takes every opportunity to build resilience while meeting immediate needs. For example, strengthening animal health in the Syrian Arab Republic through a treatment campaign that has reached 1.4 million head of sheep, cattle and goats against parasites, which has improved livestock production and health. The campaign has increased fertility rates by 15 percent and decreased livestock mortality rates by 10 percent. In drought-hit areas of the Syrian Arab Republic, FAO has worked with WFP, enabling 15 000 farmers to resume wheat cultivation, producing 30 000 tonnes of wheat, enough to meet the food needs of almost 200 000 people.

Creating and sustaining agriculture-based livelihoods not only provides food security, but it also empowers women, and builds or sustains local peace. Women are major players in the agriculture sector – they are the primary livestock caretakers, they are backyard gardeners and crop harvesters, and they play a large role in food processing. Improving agricultural productivity improves the financial and social power of these women. When better agriculture practices improve local production, equitably distribute water, and sustain rural livelihoods, they improve human wellbeing and peace.

Addressing water crises and local-level conflicts
in Yemen

The Middle East is among the most water stressed regions in the world and prospects for addressing water scarcity diminish every year due to rapidly dwindling water resources.
These resources are extremely susceptible to demand increases that come with a rapidly increasing population, damage from over-pumping, pollution and the effects of climate change. These issues disproportionately affect the most vulnerable people – women, children and the displaced.

However, better management and increased cooperation in accessing and using water could alleviate other disputes and grievances and generate benefits, including improved food security and livelihoods as well as access to energy.

In Yemen, natural resource disputes are a chronic, debilitating reality for a great many people. Violence over land and

water claims thousands of lives each year and severely inhibits social and economic development. In addition to the violence, poor natural resource management means that potentially productive land remains unused, valuable crops are destroyed, and investments are delayed or cancelled.

Land and water conflicts in Yemen have traditionally been addressed by customary regulations, however recently these mechanisms have been undermined by wide-ranging socio-economic and political changes.

FAO has been helping to resolve these conflicts and enable vulnerable families to take advantage of resources to improve sustainability and give women more opportunities to exert themselves in the country's conservative decision-making processes. For the past three years, FAO has been supporting the

establishment and reorganization of 38 Water User Associations in the Sana'a basin to better regulate water consumption and allocation. FAO has helped the associations with funding, training, and equipment.

One recent example is the successful resolution of the 17-year dispute over the Al Malaka Dam, which laid waste to nearly 170 000 m^3 of water per year – much of it lost to evaporation – enough to irrigate 34 ha of land.

In 2017, FAO started a mediation process over access to and management of water from the dam. An agreement was reached by constructing a number of shallow wells whereby the water in the dam will be stored in these wells. FAO supported the community in constructing the shallow wells through a cash-for-work programme. The wells elevate the low level of groundwater and directly benefit local agricultural production.

In reaching the agreement to end the conflict, FAO supported a more prominent role for women, by exposing them first to a comprehensive conflict resolution capacity development programme, then to ensuring their participation in the overall conflict resolution process and mediation. The role of women in raising local community awareness was instrumental in mobilizing the community members to unite behind finding a solution to the conflict.

©FAO/S. Ahmed

©FAO/A. Adil

In the Sudan, FAO is working with UNDP to support consensus-building among communities and relevant authorities on reforms to people-centred land title systems by addressing land concerns at return sites through mapping return village sites in a conflict-sensitive manner to ensure returns and reintegration processes are sustainable and conflict free. This includes demarcating nomadic corridors and revitalizing and/or establishing a fully functioning, real-time, monitoring mechanism for nomadic corridors, engaging the native administration at various levels of

the system's hierarchy. In 2018, over 100 km of livestock migratory routes were demarcated and mapped.

Ultimately, what is required is a shift in the thinking behind decision-making and investments – moving from reactive disaster and crisis management approaches toward more proactive risk management. Building agricultural livelihoods that are resilient to disasters and crises is key to securing sustainable development gains, ensuring that food and agriculture systems are productive and risk sensitive.

Advocating for climate resilience

at COP24

FAO took advantage of its participation in COP24 in Katowice, Poland in December 2018 to advocate for investment in climate resilience and greater partnership across the humanitarian-development nexus to address underlying vulnerabilities and climate-related risks to avert food crises.

A noteworthy event was the high-level roundtable, co-led by FAO as part of the Climate Resilience Network, on climate action and resilience. The roundtable focused on actions that public, private and community actors are and will be taking to accelerate investments to meet the needs of the most vulnerable people and countries to enhance climate resilience. The event emphasized that climate adaptation and resilience building are the foundation and cornerstone of sustainable development and participants developed short-, medium- and long-term recommendations for action, including:

- Committing to building resilience based on already available knowledge and successful practices;
- Leaving no one behind, specifically incorporating the needs of women, youth and people with disabilities, as well as indigenous and marginalized groups;
- Supporting community leadership and building strong local capacities;
- Promoting transformative, context- and sector-specific climate innovation and technology.

The recommendations are fully in line with FAO's resilience programme and underscore the critical need to invest in climate resilience to address rising levels of acute hunger and avert food crises in the future.

Together with the Red Cross and Red Crescent Climate Center, the International Institute for Environment and Development and other partners, FAO co-organized the Development and Climate Days 2018. Participants came together to harness their experience, evidence and learning to influence the climate negotiations, business sector planning and the formulation of climate adaptation strategies in-country.

FAO organized the Leadership Group meeting of the UN Climate Resilience Initiative that was hosted by the Development and Climate Days, which emphasized the urgency to meet the needs of the most vulnerable with coherent support by working together to scale up capacity building, knowledge sharing, innovation, finance, nature-based solutions, resilience and monitoring and evaluation.

FAO also took the opportunity to present the 2018 State of Food Security and Nutrition in the World report, which focused on building climate resilience and hosted an interactive panel discussion on localizing climate resilience action to meet the needs of Indigenous People.

©FAO/A. Proto

FAO is therefore also working with governments and regional organizations to ensure an enabling environment for resilience building. For example, in 2018, FAO helped build the technical capacities of countries in Latin America and the Caribbean (Chile, Colombia, Peru and some Caribbean small-island developing states) to report damages and losses in the agriculture sector caused by disasters. This is crucial, not just to enable them to comply with reporting requirements against the Sustainable Development Goal and Sendai targets, but also to ensure policy- and decision-makers have the highest quality information to determine actions and policies to reduce the effects of disasters.

In addition, a regional governance mechanism was established to implement the Community of Latin American States Regional Strategy for Disaster Risk Management in the Agriculture and Food Security and Nutrition Sectors (2018–2030). Through its support to the establishment of an Executive Committee and Advisory Group, FAO fostered collaboration among countries, subregional organizations, research institutions, resource partners and UN agencies, enabling the development of proposals and mobilizing resources to implement the Strategy and reduce losses to disasters.

Women's empowerment is critical for the development of rural communities. Women also experience very specific challenges in crisis situations; for example, they often have less access to, and rights over, land, water and training opportunities. In contexts like the Syrian Arab Republic, FAO has placed a significant emphasis on supporting women and improving the nutrition of children, recognizing that eight years of conflict have severely undermined their food security and nutrition and left them highly reliant on humanitarian assistance for their survival. This has included vegetable production and training to increase women's capacity in marketing their products, protecting small herders' assets through the national animal health campaign, as well as school gardening projects that increase access to vegetables and nutritious foods in schools.

©FAO/J. Almere

Supporting women to enhance food production
in the Syrian Arab Republic

Increasing women's involvement in agriculture and food production, will contribute to improving the availability of food in local markets, providing women with more opportunities and helping to form stronger and more stable communities when women are able to meet their own needs and those of their children, including for schooling and health care.

In 2018, FAO reached more than 12 100 women-headed households with agricultural inputs accompanied by intensive technical training. Rural women have received production kits such as seasonal vegetables seeds, modern irrigation equipment, poultry production inputs and many more.

Mouna Yasser, a 30-year-old mother of two from Homs Governorate, had been entirely dependent on her husband's salary until they lost their only source of income. Through FAO she received 45 laying hens, five roosters and training on poultry keeping. "I hope I can raise more hens to be able to sell them. I feed my chickens very well twice a day to produce more eggs; I will be able to sell the eggs to the market to support my family with good income." She added: "I am now taking care of my children and feed them with daily fresh eggs."

FAO is expanding this support, based on the country's needs. A concentrated training programme, entitled "The Road to the Market" has provided more than 80 rural women with skills in food processing, branding and business to enable their goods to compete in the local market.

In addition, FAO is raising awareness of the nutritious value of vegetables and other food, and enhancing the diets, of more than 3 900 schoolchildren, teachers and their local communities, through a joint United Nations project, entitled "Education for All". The intervention has encouraged children to learn more about vegetable production through setting up school gardens in each of the targeted schools. They have planted tomatoes, cucumber, green peppers, eggplants and more. The children were particularly excited to harvest their crops, taste their vegetables and their experiences with their family and friends. This approach also acts as an entry point to reach children's families and communities in order to mainstream the consumption of nutrition-rich food to address hunger and malnutrition.

Hunger and conflict have been key drivers of a dramatic increase in forced migration in recent decades from 40 million in 1997 to 68.5 million people in 2017. Countries with the highest rates of refugee outflows are from countries experiencing both armed conflict, and high levels of food insecurity and hunger. As displacement becomes ever more protracted, FAO is working to provide displaced people and their hosts with opportunities to rebuild their lives and move from 'care and maintenance' to 'inclusion and self-reliance'.

This means providing real livelihood opportunities for both displaced populations and their host communities, who are often themselves facing considerable challenges. Without this, the presence of growing numbers of displaced can fuel tensions through competition for basic services and scarce natural resources – accelerated by environmental degradation and climate change.

Investing in resilient livelihoods is key. In many of the countries that account for the majority of the world's displaced people (Afghanistan, the Democratic Republic of the Congo, South Sudan, Somalia and the Syrian Arab Republic), the majority of people relied on agriculture-based livelihoods before their displacement.

While the protracted nature of displacement is of real concern for those displaced and their livelihoods, the lengthy nature of displacement does create opportunities to enhance local economies and bring skills, capital and connectivity to broader markets that might fill the unmet needs of the host communities.

Over the longer term, activities need to be geared more towards creating conditions that are more conducive to building self-sufficiency while people remain in displacement. These can strengthen self-reliance and help people seize opportunities, as soon as they appear, paving the way for durable solutions.

In these contexts, interventions that strengthen local participation in decision-making processes on natural resource management and use are vital and can provide technical entry points to engage different groups and to foster the emergence of inclusive governance systems. Such interventions can strengthen social cohesion and increase peace dividends amongst and between IDPs and host communities, and for returnees when they can go home.

Addressing conflict and hunger, the interaction between the two and other compounding factors such as climate shocks and variability, is therefore critical to easing forced migration pressures in these contexts. Investment in resilient, agriculture-based livelihoods and strong food systems, can not only help to address some of the drivers of forced displacement, but also

enhance social cohesion, support host communities and contribute to the social and economic integration of vulnerable populations, including displaced, migrant communities.

This starts from evidence. For example, using the Resilience Index Measurement and Analysis (RIMA) to gain a better insight of the capacities of host communities as well as displaced populations. For host communities, the influx of large numbers of forcibly displaced disrupts existing equilibria, straining available resources and creating mismatches in demand and supply in multiple markets. The inflow of migrants often calls for a redefinition of priorities in hosting countries and external assistance. Measurement of the resilience of both host and displaced populations can provide a solid base to inform responsive measures to lessen the negative impact of displacement and harness its development potential.

FAO has supported the use of RIMA by national authorities in parts of Uganda that are risk-prone as a result of the arrival of more than 1 million South Sudanese refugees. The RIMA allows a better understanding of the dynamics between refugee and host communities and will facilitate a more needs-based allocation of funding for concurrent humanitarian, development and peace interventions.

FAO has been working to enhance social cohesion and stability and strengthen the local economy through investment in the restoration of livelihoods of both host and displaced communities, through supporting local entrepreneurship and local level employment creation. FAO is taking this approach in Burundi where a significant number of refugees are now choosing to returning home, placing great pressure on an already fragile country and local communities who are dependent on humanitarian assistance for survival.

Returnees are facing great socio-economic reintegration challenges, including, in some cases acute food security, and loss of land and livelihoods. In Burundi, FAO is working together with local partners, receiving communities and returnees to support the transition of refugees and returnees from aid dependence towards resilience and self-reliance, building their capacity to contribute to not only their own socio-economic recovery, but also the economic development of their communities.

Such interventions can create a pathway to a more prosperous, peaceful and stable future for both host and displaced communities. Engaging refugees, returnees and host communities in local economies, not only makes good sense developmentally, but also economically.

Risk-informed and shock-responsive social protection

Disasters and crises not only exacerbate the existing vulnerabilities of rural people dependent on natural resources, but can also reverse years of progress in terms of hunger and poverty reduction. Well-designed social protection programmes can not only help address the structural causes of chronic poverty and enhance the productive capacity of the poorest, but can also ease humanitarian caseloads in crises.

FAO recognizes the critical importance of identifying innovative and scalable approaches to help reverse the trend of rising hunger. Risk-informed and shock-responsive social protection systems provide the humanitarian community with well-developed structures at scale to effectively respond to crises – supporting the poorest and the most vulnerable. In addition, the extensive operational expertise of humanitarian actors in fragile contexts can provide the development community with innovative approaches, supporting the design of ad hoc social protection systems to address the varying needs of people affected by cyclical threats and crises.

Together with partners such as UNICEF, WFP, the World Bank and others, FAO has identified "risk-informed and shock-responsive social protection" as a key innovation in supporting the poorest and most vulnerable in society.

©FAO/V. Villafranca

In this regard, in 2018, FAO supported countries in the design of risk-informed shock-responsive social protection, across three dimensions:

- In contexts where social protection programmes or systems exist, work with key counterparts to enhance the linkages with livelihoods support, as well as working to make systems more responsive.

  FAO in partnership with the International Labour Organization (ILO), UNICEF, WFP and the United Nations Office for Disaster Risk Reduction, is developing a joint project to strengthen capacity of Association of Southeast Asian Nations (ASEAN) member states (AMS) to develop risk-informed and shock-responsive social protection systems. The specific objective is to improve availability of policy and operational options for AMS to strengthen shock-responsiveness of social protection systems including reference to design options, financing and scale-up triggers. Main activities include (i) an ASEAN-wide study of existing social protection systems and policy options, (ii) in-depth country analysis in Cambodia, Myanmar, Viet Nam and the Philippines, (iii) country roadmaps in selected AMS to strengthen early warning systems to trigger shock-responsive social protection, and (iv) ASEAN guidelines on risk-informed and shock-responsive social protection.

- In contexts where no systems are in place or where innovation is needed, FAO has been designing and implementing cash-based interventions, including cash+.

  FAO has developed a strong expertise in the implementation of cash-based programmes linked to livelihoods promotion and agricultural development, especially in fragile and protracted crisis contexts. Moreover, FAO has been supporting the design of nascent social protection systems based on the experience and lessons learned from the implementation of humanitarian cash-based operations. As an example, in close partnership with WFP and UNICEF, FAO has been supporting the development of a social protection strategy in Somalia.

FAO has been also working to enhance operational knowledge on the design of risk-informed and shock-responsive social protection systems. FAO is currently developing internal guidance to integrate social protection as one pillar for resilience programming, particularly in contexts of fragility and forced displacement. Moreover, in partnership with WFP, UNICEF and others, FAO has supported the development of analytical reviews to assess the potential entry point to strengthen linkages between humanitarian and developing programming through social protection in Afghanistan, Somalia and Bangladesh. Similarly, FAO will support the design of livelihood interventions for refugees and host communities in Lebanon, Turkey and Jordan, as well as explore their potential integration into existing social assistance processes.

In the context of guidance and knowledge generation, FAO partnered with the Red Cross Red Crescent Climate Centre to develop a framework document and capacity development materials in the linkages between social protection and climate-related risks.

- In contexts of forced displacement, FAO works to support the livelihood of refugees, as well as of host communities, particularly those living in rural areas. For example, aligned with the Lebanon Crisis Response Plan 2017–2020, FAO contributes to the current response to the Syrian crisis in Lebanon by providing technical assistance to the Government of Lebanon to expand the coverage of social protection to farmers, fishers and workers in those sectors; deliver work permits for Syrian workers in agriculture with the ILO, and; establish coordination mechanisms between the Ministry of Agriculture and the Ministry of Social Affairs. On the other hand, in collaboration with IFAD and WFP, FAO provides direct support to refugees and host communities in rural areas to promote their agricultural livelihood, encouraging the overall stabilization and social cohesion in Lebanon.

Strengthening pastoral livelihoods, focusing on Africa's drylands

Climatic variability and extremes are squeezing Africa's livestock owners, compounding existing high levels of vulnerability linked to decades of marginalization, environmental degradation, population growth, instability and recurrent shocks. Shrinking rangelands and increasing water scarcity are driving destitution and rising acute hunger among pastoral and agropastoral populations.

Despite their weakening capacity, pastoral communities remain highly resilient and make enormous contributions to social, environmental and economic wellbeing in the dryland areas. FAO is thus strengthening the resilience of these communities, through a combination of interventions that build capacity to adapt to changing circumstances.

FAO's support includes improving capacity and accountability in governance institutions, addressing cross-border and regional aspects of pastoralism, using monitoring systems to address problems when they arise and ensuring a timely livelihoods-based livestock emergency response when crises threaten.

The cross-border and regional dimension of pastoralism require that policies that affect pastoral communities and their livelihoods are harmonized across regions, with strong cooperation among states. In this, for example, FAO works closely with the Intergovernmental Authority on Development (IGAD) to foster regional approaches to pastoralism in the Horn of Africa, including supporting the development of an Animal Feed Action Plan for Eastern Africa.

More on pastoralism in Africa's drylands

©FAO/A. G.Farran

Predictive Livestock Early Warning System (PLEWS)

PLEWS is a tool developed by FAO and Texas A&M University, which predicts edible vegetation and surface water availability for livestock herds. PLEWS has been used to demonstrate the direct relationship between forage availability and human malnutrition in Kenya and is a crucial tool for informing early warning and action in response to deteriorating fodder conditions. PLEWS also plays a critical role in early warning, for example, not just demonstrating where feed gaps are likely but where transhumant populations are likely to move with their livestock, and therefore where potential conflicts could emerge, and where we need to intervene in terms of service provision like water sources (e.g. bladder water tanks) and veterinary support.

Investing in research to build evidence and inform action and advocacy

FAO is convinced that research and learning is crucial to constantly re-examine our assumptions and provide a closer understanding of the needs of communities and inter-relationship between different challenges they face if we are to achieve our goal of ending hunger and avoid deepening vulnerabilities and tensions at the community level.

For example, previous FAO-led studies in some pastoral communities in Kenya have shown that there is a direct correlation between rising child malnutrition and deteriorating availability of fodder and water for livestock. In 2018, to better understand this linkage and inform more appropriate interventions for pastoral communities, FAO and UNICEF, together with Washington State University, Kenya's National Drought Management Agency and Marsabit county government, further supported research on the interconnectedness between children's health and animal health. The project is using a cluster randomized control trial providing livestock feed, and livestock feed with nutritional training to targeted pastoral households during critical dry periods and comparing this to a control group of households. This is essentially enabling monitoring of the efficiency and effectiveness

of these types of interventions in curtailing spikes in malnutrition. The ultimate goal is to develop a replicable and scalable strategy to protect pastoral families against seasonal rises in acute malnutrition for children and pregnant women during drought periods.

Linked to this research is FAO's advocacy efforts, including preparing a note for the United Nations Executive Committee on farmer-herder conflicts in West Africa and the Sahel in 2018, a policy brief on farmer-herder conflicts in the region and organizing a workshop with Tufts Univeristy, entitled "Mind the gap - bridging the research, practice and policy divide to enhance livelihoods resilience in conflict settings".

Feed balance

In 2018, FAO placed a new emphasis on supporting stakeholders in the Horn of Africa to evaluate feed availability at the national and local levels, not just during disasters. Given the reality of increasing climate variability and extremes in the region – characterized in particular by recurrent droughts and reduced periods between severe droughts – there is a clear need to be more strategic in terms of feed access and availability for livestock herds.

In 2018, FAO and IGAD undertook a multistakeholder consultation with member states, the private sector, development partners and community institutions to address livestock feed challenges in the region. Based on these, the two organizations have developed an Animal Feed Action Plan for Eastern Africa released in early 2019.

FAO has developed the feed balance sheet to analyse feed availability in a given country or region based on domestic availability of resources. This is critical for strategic planning, including for early warning and early action, providing real-time data as well as forecasting changes in the coming months to inform pre-positioning of feed reserves or other actions to protect livestock herds against feed deficits.

In Kenya, between May and June 2018, FAO delivered a number of capacity development activities around livestock feed inventory as a means of analysing data on

©FAO/M. Tewe

livestock feed in terms of type, quantity and available balance. These trainings targeted 23 county government staff from arid and semi-arid land areas and a further 17 staff from the national governments, private sector, research and higher learning institutions. To further strengthen capacity for analysis of the feed balance, all participants were also trained on use of PLEWS, specifically on estimating feed balance and availability.

In Ethiopia, FAO worked closely with the Government and relevant institutions to develop a feed inventory and balance for the country. Information generated through the study is assisting the Government, resource partners and private sector actors to formulate investment strategies for developing the livestock feed sector (at commercial and community levels), as well as enabling them to address and effectively manage feed shortages during droughts. A pre-requisite for making the best use of available feed resources is to accurately assess and understand their availability, location and nutritive value.

FAO also supported a study to determine the availability of agro-industrial byproducts and their use as animal feed in Ethiopia. Given that the poor quality and inadequate quantity of livestock feed is a major challenge to livestock production in the

region (and beyond), these byproducts could provide an importance source of additional and nutritious feed. To-date, their use as feed in Ethiopia is almost negligible. The study is thus also contributing to enhance awareness of these byproducts as a source of animal feed and build capacity to properly manage them at production sites and on-farm. Additional feed sources identified could also help reduce the national feed balance gap.

©FAO/M. Tewe

Actively strengthening pastoral livelihoods

FAO is also engaged in strengthening pastoral livelihoods at the community level, for example through pastoral/agropastoral field schools in countries including South Sudan, Kenya, the Niger, Mali to enhance livestock production, trade and peace initiatives, especially in cross-border areas.

More on pastoral field schools (PFS)

In Mandera county of northeastern Kenya, pastoralists are receiving assistance to grow pasture for their livestock, strengthening their resilience in times of drought. Mandera is in Kenya's arid and semi-arid lands and part of the cross-border region between Ethiopia, Kenya and Somalia, where pastoralists are particularly vulnerable to food insecurity. Through an IGAD-FAO partnership programme on drought resilience, agropastoral field schools have been set up to support pastoralists in producing, managing and using fodder. Women, men and youth are part of the schools, helping break down traditional barriers to women's participation in these tasks.

"With the agropastoral field schools, women are able to produce pasture bale and store it so that they are no longer dependent on men," said Shanqaray Hassan Mohamed, the vice chair of one of the agropastoral field school groups. "We have successfully replicated the fodder production at our farms, improving our production".

In the Sudan, too, FAO has continued to support pastoral livelihoods including through the construction of 600 km of firebreak lines to protect rangeland and pastures, providing almost 7 tonnes of local palatable pasture seeds for pasture rehabilitation, and the rehabilitation of two water storage hafirs (reservoirs) in West Darfur State.

Responding to the pastoral crisis in partnership with UNICEF and WFP
in the Sahel

The 2017/18 pastoral season in the Sahel was marked by a deficit and poor distribution of rains, which resulted in significant biomass shortfalls and early drying up of water points, significantly affecting pastoral livelihoods.

To mitigate the impact of the crisis and strengthen the resilience of affected people, FAO developed an emergency response programme for the six affected countries (Burkina Faso, Chad, Mali, Mauritania, the Niger and Senegal), which was coordinated by the FAO Regional Office for Resilience in West Africa and the Sahel.

The programme reached almost 1.4 million people in the region, through a combination of interventions: livestock feed, animal health support, veterinary kits, cash transfers, seeds and tools, construction or rehabilitation of water points, destocking, solar pumps, and training.

Over 12.5 tonnes of livestock feed were distributed, which had a number of positive effects, including significantly increasing the amount of milk produced by animals each day, which contributes to combating malnutrition among children. In addition, livestock were overweight during the lean season, enabling benefiting families to sell them for twice the price of other animals, and resulting in an increase in births among herds. Vulnerable families have thus increased their incomes, which they have invested in diversifying their diets, medical care, clothing, school fees and supplies. Critically, by protecting core breeding herds, beneficiaries were able to reduce distress sales of livestock and maintain their herds.

"I received the nutritional blocks and cakes for my animals (sheep). In two months, I found that my animals were overweight and that there was an increase in milk production of about 0.5 liters per day per female. This increase allowed me to have plenty of milk for my children and other household members. I was able to sell two goats to meet the family needs, and as these animals were well fed, I earned XAD 32 500 (USD 55.40). That money allowed me to buy food for the household, the straw stock for the other animals, soap, and so on. Due to the positive effects of the meal, I decided to buy corn bran at the market for my animals in case the endowment would be exhausted" – recipient of FAO-provided animal feed in Chad.

In addition, the seeds distributed – particularly the pigeon peas – were critical for feeding livestock, while the provision of small ruminants, vegetable seeds and agricultural tools was particularly beneficial during the lean season, ensuring a source of food and income during crisis.

Preventing, detecting and responding to animal disease threats

Many communities rely on animals for their livelihoods as well as their food security and nutrition. When animal diseases jump from animals to humans – like recent pandemics such as avian influenza and Ebola – they can spread around the world in a matter of hours or days, posing a threat to global health security and the livelihoods of livestock farmers.

FAO is building animal health capacity to prevent, detect and respond to disease threats in more than 120 countries. Through its Emergency Centre for Transboundary Animal Diseases (ECTAD), FAO plans and delivers animal health assistance to member countries responding to animal disease threats – working to reduce the impact of animal diseases on lives and livelihoods, and helping to stop emergence and spread of potential pandemics at source.

ECTAD multidisciplinary teams facilitate two key animal health programmes funded by the United States Agency for International Development: the Global Health Security Agenda (GHSA) and Emerging Pandemic Threats (EPT). Building animal health capacity, these two programmes are reducing the risk of national, regional and global disease spread.

Using a multisectoral One Health approach, the GHSA programme develops national capacity to prevent zoonotic and non-zoonotic diseases while quickly and effectively detecting and controlling diseases when they do emerge. The EPT programme focuses on improving national capacity to pre-empt the emergence of infectious zoonotic diseases and prevent pandemics.

Global achievements (October 2017–September 2018)

 Responded to
190 outbreaks in 18 countries, by providing support including: vaccination, outbreak investigation, technical advice and coordination.

 Trained more than
7 100 professionals, increasing preparedness by building capacity to prevent, detect and respond to infectious disease outbreaks.

 Helped to improve the diagnostic capacity of
66 laboratories in 23 countries. Reducing time taken from field sampling to accurate diagnosis of priority zoonotic diseases leads to faster and more effective response.

©FAO/Egypt

©FAO/H. D. Nam

Through ECTAD, FAO is building animal health capacity in over 30 countries

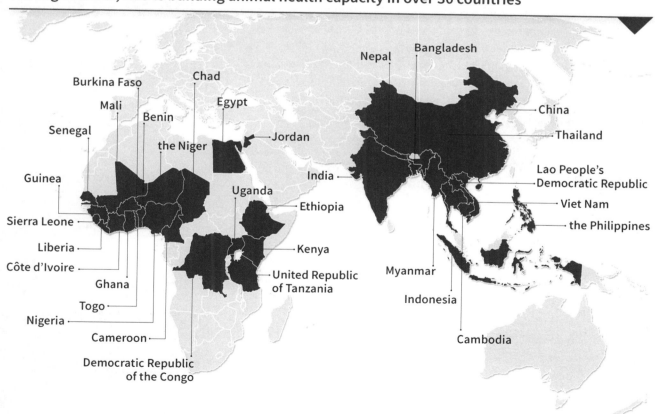

Burkina Faso

Mali

Senegal

Chad

Benin

Egypt

the Niger

Guinea

Jordan

Sierra Leone

Uganda

India

Liberia

Ethiopia

Côte d'Ivoire

Kenya

Ghana

United Republic
of Tanzania

Togo

Nigeria

Cameroon

Democratic Republic
of the Congo

Nepal

Bangladesh

China

Thailand

Lao People's
Democratic Republic

Viet Nam

the Philippines

Myanmar

Indonesia

Cambodia

Conflict and peace

In 2018 there was even greater acceptance and recognition that conflict and hunger reinforce each other and that conflict is a key driver of situations of acute food insecurity. The adoption of United Nations Security Council Resolution 2417 (UNSCR 2417) on the *Protection of civilians in armed conflict* in May 2018, considers conflict-induced food insecurity, including famine, as a threat to international peace and security, and highlights the link between conflict and hunger. Among other things, UNSCR 2417 points to the need for early action linked to early warning, and to invest in resilience by safeguarding agriculture-based livelihoods as an essential contribution towards preventing and responding to food crises, and supporting local stability.

Armed conflict (both directly and indirectly) affects food security and nutrition, and contributes significantly to forced displacement, while also serving to impede humanitarian response, thereby increasing the risk of severe food insecurity and famine. Some of the root causes of hunger (including poverty, inequality, lack of access to natural and productive resources, forced displacement, and climate change impacts) can contribute to and exacerbate conflicts.

In March 2018, FAO published its *Corporate Framework to Support Sustainable Peace in the Context of Agenda 2030* – the intent being to improve how FAO (i) mitigates the negative impacts of conflicts on people's lives and livelihoods, (ii) prevents the risk of local conflicts, whilst (iii) promoting a transformative agenda to address the root causes of local conflicts and promote sustainable development. Building and strengthening resilience to conflict requires helping countries and households prevent, anticipate, prepare for, cope with, and recover from conflicts.

More on FAO's framework to support sustainable peace

Equally, the importance of investing in resilient agricultural livelihoods as a key contributor to stability and local peace should not be underestimated. FAO recognizes that ensuring that interventions do not heighten conflict risks and avoid doing harm, as well as identifying where opportunities may exist to contribute to local peace, needs robust contextual understanding.

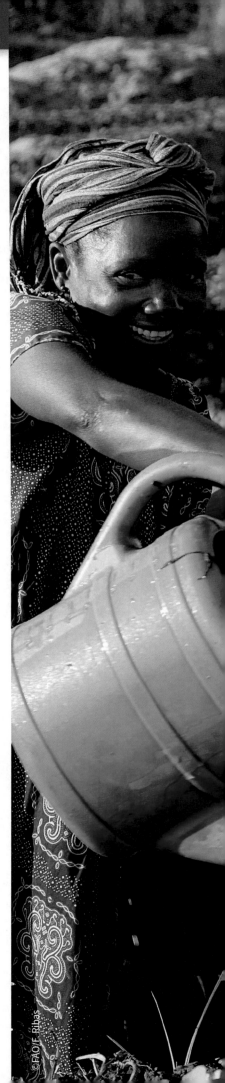

FAO's interventions in fragile and conflict-affected contexts

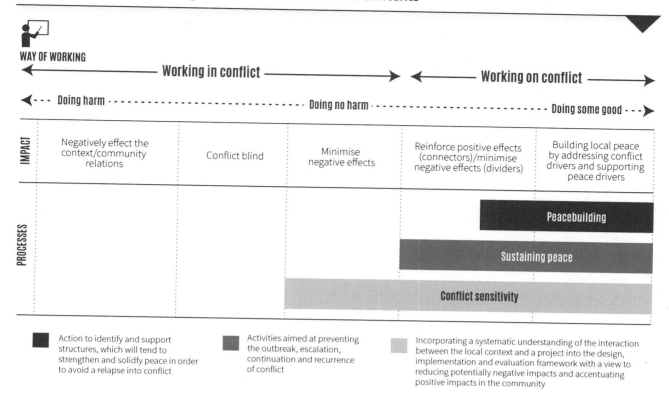

A sustainable impact on peace is more likely when food security and livelihood-related initiatives are implemented as part of a broader set of multisectoral, humanitarian, developmental and peace-related interventions.

In this regard, FAO works across the humanitarian-development-peace nexus by focusing on building resilience – which links humanitarian and development action – while at the same time identifying opportunities to contribute to improved prospects for local peace. This can include improving natural resource management between identity groups, or supporting social cohesion through agricultural activities that increase contact between people.

Through a number of new partnerships – for example with Interpeace and Uppsala University – FAO improved its capacities and generated new evidence to inform programmes and policies, so that we can do what we already do, but better.

Exploring the links between resilience, peace and conflict
in the Democratic Republic of the Congo

The effects of climate variability and change on security are hotly debated. While this topic has received considerable attention in both policy circles and academia in recent years, understanding of the conditions under which climatic shocks affect armed conflict are still limited and even in high-risk regions there is great variation in who will regard the use of violence as a viable option in response to a climate-related shock. In partnership with Uppsala University, household survey data from a multi-year resilience programme funded by the Government of Canada in North Kivu has been analysed to study violent attitudes following agricultural production shocks as a result of climatic change. Preliminary findings indicate that less resilient respondents, based on both objective and subjective indicators, are more likely to be supportive of violence following agricultural production shocks.

This is just a start in building the evidence base on these relationships, but such findings tentatively support the hypothesis that increasing resilience can contribute to localized stability and improved prospects for local peace.

Ensuring conflict-sensitivity in programming

In 2018, FAO partnered with Interpeace to develop FAO-specific tools, guidance and training on conflict-sensitivity and context analysis. This partnership brings together FAO's technical and programmatic knowledge with Interpeace's 25 years of experience in peacebuilding.

A critical tool developed and field-tested with Interpeace in 2018 is the Conflict-sensitive Programme Clinic, a structured participatory analysis designed to identify and integrate conflict-sensitive strategies into the design and implementation of FAO interventions. In 2018, support on this was provided to FAO Country Offices in Iraq, Jordan, Kenya, Lebanon, Mauritania, Nigeria, Palestine, Somalia, the Sudan, the Syrian Arab Republic and Turkey, as well as the Regional Office in Cairo and the SP5 Resilience Team in East Africa, and informed programme development, including the Global Programme against Food Crises and the United Nations Peacebuilding Fund (PBF) projects.

Testimonies from the Conflict-sensitive Programme Clinic, Nairobi, 5-6 December 2018

"The approach is great, the main reason being that it really is learning by doing. It is not extremely complicated to do, nor is it extremely time consuming and the process is approachable for non-experts."

Cyril Ferrand, Team Leader,
FAO Resilience Team in East Africa

"What I like about this particular process, is that in a very simple way it allows someone to put in a lot of complex issues and get a strong result. The simplicity of it is in fact my main take home message. You can follow this process and come up with very concrete conflict-sensitive recommendations factoring in many different elements."

Michael Gitonga, Land, Water and
Irrigation Officer, FAO Kenya

Contributing to peacebuilding through the PBF

FAO has considerably stepped up its engagement with the PBF over the past two years, and the high success rate of project approvals in 2018 represent concrete recognition by the UN Peacebuilding Support Office of FAO's role in implementing activities that contribute to local peace, particularly over natural resource management.

In 2018, 16 PBF projects with FAO participation were approved, with FAO as the lead agency for seven projects, including two cross-border projects in the Sahel. This represents a doubling of funding from 2017, with an 80 percent approval rate for FAO submissions to the PBF. FAO partners with a number of other United Nations agencies, funds and programmes in implementing PBF projects, including UNDP, UNICEF, WFP, UN Women, United Nations Population Fund, UNHCR, IOM and ILO.

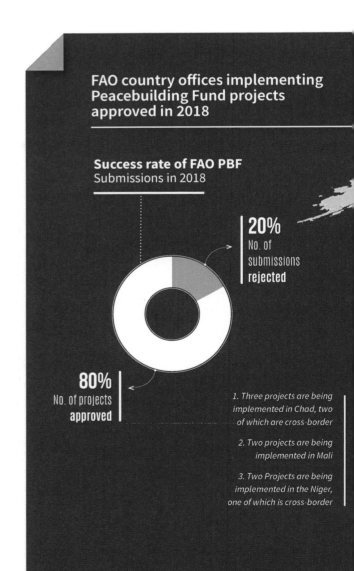

FAO country offices implementing Peacebuilding Fund projects approved in 2018

Success rate of FAO PBF Submissions in 2018

20%
No. of submissions rejected

80%
No. of projects approved

1. Three projects are being implemented in Chad, two of which are cross-border

2. Two projects are being implemented in Mali

3. Two Projects are being implemented in the Niger, one of which is cross-border

FAO's mandate and core technical capacities are highly conducive to improving the prospects for local peace. Thematic areas of FAO interventions under current PBF projects include:

- *Natural Resource Management*: Promoting participatory and inclusive management of natural resources in support of peaceful dispute resolution, improved social cohesion and longer-term sustainability

- *Rural Livelihoods Support*: Promoting intra- and inter-community livelihoods interventions aimed at improving community cohesion and cooperation and strengthening community resilience

- *Rural Employment*: Promoting rural labour opportunities and economic autonomy through capacity development and entrepreneurialism, with a particular focus on women and youth

- *Reintegration*: Promoting rehabilitation of inter-community infrastructure and engagement mechanisms in support of post-conflict socio-economic reintegration and reconciliation of communities

- *Host-IDP relations*: Promoting joint host community-IDP programming that improves communication across and between communities, provides opportunities for inter-community economic gain, and seeks to resolve localized disputes over access to/management of natural resources, through mediation

- *Cross-Border Programming*: Supporting activities that seek to convene formal and traditional stakeholders, complemented by targeted community interventions such as water points and which, in the context of weak or absent governance and pressure on natural resource management, can work as a catalyst for longer term local peace.

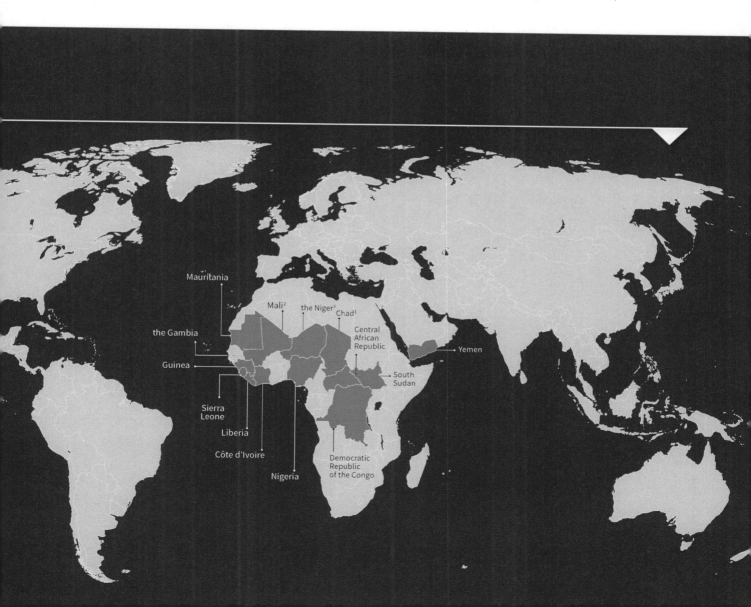

Pastoralists are key actors and an integral part of a lasting solution to ensure the stability of Africa's drylands.

Mitigating and preventing pastoralist conflict

The cross-border pastoralist communities of Kenya and Uganda have been conflict hotspots for many years. The conflicts are mainly linked to cattle raiding among the Pokot tribe in Kenya and the Karamojong in Uganda. This has been compounded by consecutive years of drought. During 2018, FAO has been strengthening the capacity of pastoral communities most vulnerable to drought by setting up PFSs not only as a way to help reduce and prevent inter-community conflicts, but also as a means to promote a learning environment where community members exchange information and best practices and learn about grassroots ways of coping with drought risks and related challenges.

Learning to live together

FAO's Dimitra Clubs are a community-based approach that builds relationships and contributes to social cohesion, improving aspirations, confidence and trust. In the Tanganyika province of the Democratic Republic of the Congo violence between Bantu and Twa has worsened in recent years. FAO and WFP are working together to increase smallholder farmers' incomes and build resilient livelihoods, but also to support community-based organizations. Over 140 Dimitra Clubs have been created, improving social cohesion between Bantu and Twa as a result of inter-group dialogue, collective action and awareness raising activities, particularly over how to manage local natural resource-based conflicts.

"We are very satisfied with the Dimitra Club initiative that has shown us how to get along together, to live with the Bantu in order to jointly work to develop our country. We live from hunting and the Bantu practice agriculture. With the Dimitra Clubs we no longer believe we should fight the Bantu, but rather we need to develop together with them."

Twa representative

Working together in 2019

"The cornerstone of the fight against hunger must be increasing the resilience of rural livelihoods to conflicts and the impacts of climate change."

Jose Graziano Da Silva, FAO Director-General

2018 saw significant progress in making collaboration across the humanitarian-development-peace nexus a reality on the ground, particularly in the form of the nascent Global Network Against Food Crises. However, given persistently high levels of acute hunger – above 100 million people for the last three years – it is clear that much more work is needed to address hunger at its roots.

The annual Global Report on Food Crises, and the 2017 and 2018 editions of the State of Food Security and Nutrition in the World provide the critical evidence of interconnectedness across the humanitarian-development-peace nexus and the importance of bringing all actors together to avert food crises in the future.

These reports make it clear that we are failing to adequately address the underlying fragilities that undermine and erode livelihoods and push people to the brink, like climate change, population growth, environmental degradation, political marginalization, growing inequality and rural poverty. The evidence tells us that we can do more to align humanitarian interventions with preventative, resilience-oriented development actions to address immediate needs and tackle underlying causes of hunger, malnutrition and vulnerability.

Therefore, it is absolutely critical that we collaborate more effectively across sectors using a systems approach, to address the multiple drivers of fragility and vulnerability through our combined work on peace and security, human rights, social protection, gender equality, equitable access to land, education, energy, environment (soil, forestry, biodiversity, etc.), climate change, water and sanitation, in order to deliver a more inclusive, holistic, equitable, resilient, and sustainable set of context-specific solutions and responses.

One example of this collaboration is the United Nations/World Bank Famine Action Mechanism, in which FAO has been a partner since its initial phase, which seeks to further enhance early warning systems and analyses (like the IPC) by improving the capacity to forecast areas at risk of famine.

Against the plethora of tools for food security, nutrition and resilience, knowledge management can promote the harmonization across approaches and foster coordination and coherence. The FAO-led knowledge sharing platform for resilience, KORE (Knowledge Resilience), offers a structured space to generate, capture and exchange knowledge between FAO and other key partners. Through activities such as webinar

series, good practices capture and methodological support for knowledge generation, KORE works with key development actors and fosters a culture of learning to strengthen resilience to food insecurity and malnutrition in the face of shocks and stresses.

At the same time, FAO and our partners will continue to advocate for the observance of and compliance with international law and international humanitarian law – as laid out in United Nations Security Council Resolution 2417 – by all relevant actors to reduce the impact of conflicts on food security.

In 2019, within the strategic programme on resilience, FAO will therefore focus on further operationalizing the Global Network Against Food Crises and upscaling its interventions across the humanitarian-development-peace nexus through the Global Network's membership, which has expanded to 15 partners since 2017.

The Global Network Against Food Crises

The Global Network Against Food Crises was launched at the 2016 World Humanitarian Summit by the European Union, FAO and WFP with the objective of tackling the root causes of food crises through shared analysis and knowledge and strengthened coordination in evidence-based responses across the humanitarian-development nexus.

The Global Network acknowledges the centrality of food and agri-food systems in preventing food crises and mitigating their impact, boosting recovery and reconstruction. It also acknowledges the need to understand links and coordinate policies and actions in relation to other complex dynamics and drivers of vulnerability, such as conflict and insecurity, climate change and demography.

It is a global platform aiming to shape food security and nutrition decision-making by establishing and consolidating partnerships at national, regional and global levels; sharing data and analyses; defining innovative approaches; monitoring progress towards better food security; pursuing evidence-based advocacy; and coordinating support for food security and nutrition in contexts at risk of food crises within a longer-term perspective of eradicating hunger and malnutrition by 2030.

©FAO/J.Jadin

In particular, FAO in collaboration with the other partners will work through a series of actions in order to operationalize the three interlinked dimensions and levels of the Global Network:

- Generate evidence-based information and analysis – at country, regional and global levels – through the flow of information for the preparation of the annual Global Report on Food Crises and related products and updates, including real-time Web-based updates, complemented by early warning information.

- Leverage strategic investments to prevent and respond to food crises along the humanitarian-development nexus – at country, regional and global levels – through action plans at country and regional levels that feed the preparation of the annual Global Report on Food Crises' Prevention. Given the aspiration of the Global Network to promote the operationalization of the humanitarian-development-peace nexus in the area of food crises and food systems' transformation, the Global Report on Food Crises' Prevention will include analysis of the relevant response actions and policies at national, regional and global levels, as well as their impact. This will contribute to improve information sharing on planned interventions at the three levels and in developing a monitoring system under the framework of the Global Network, to regularly assess mutual progress in effectively working together to achieve collective outcomes.

- Foster political uptake of information generated through the other two dimensions to formulate strategic policy orientations aimed at coordinating operational-level actors in implementing programmes and investments at regional and country levels. This will facilitate higher-level political uptake and enhanced coordination with other actors such as peace, environmental, climate and other relevant stakeholders to more effectively prevent and respond to food crises.

Partnerships like the Global Network allow FAO to leverage its expertise and maximize its impact in addressing the root causes of food crises at the agriculture and food systems' level.

FAO will thus continue to boost the Global Network's work by engaging with other initiatives and teaming up with a range of partners – within the United Nations system and beyond, from the global level down to neighborhoods and communities – to explore new approaches to preventing and responding to food crises and breaking the cycle of vulnerability.

No longer can humanitarian efforts be focused only on disaster relief. A new paradigm has emerged, one that emphasizes reducing risks ahead of time and supporting risk-informed development that enhances resilience to shocks and enables swifter, more sustainable recoveries.

Weakness in resilience can trigger a downward spiral after crises hit – on a very human scale, when communities' livelihoods are wiped away – but also at the national or larger levels, as development gains that took years to attain can be compromised and lost.

But resilient agriculture is not just a mere means of subsistence or survival; rather, it lies at the heart of sustainable development.

This is why FAO has firmly positioned building the resilience of rural communities front and centre in its ongoing work to create a world with Zero Hunger.

"The Global Network Against Food Crises represents a key step in providing better understanding of a crisis, facilitating integrated and agreed upon programme options, and advocating at the highest levels for collaborative action to prevent food crises."

Daniel J. Gustafson, FAO Deputy Director-General, Programmes